This report contains the collective views of an international group of experts and does not necessarily represent the decisions or the stated policy of the United Nations Environment Programme, the International Labour Organisation, or the World Health Organization.

**Environmental Health Criteria 155**

# BIOMARKERS AND RISK ASSESSMENT: CONCEPTS AND PRINCIPLES

Published under the joint sponsorship of
the United Nations Environment Programme,
the International Labour Organisation,
and the World Health Organization

World Health Organization
Geneva, 1993

The **International Programme on Chemical Safety (IPCS)** is a joint venture of the United Nations Environment Programme, the International Labour Organisation, and the World Health Organization. The main objective of the IPCS is to carry out and disseminate evaluations of the effects of chemicals on human health and the quality of the environment. Supporting activities include the development of epidemiological, experimental laboratory, and risk-assessment methods that could produce internationally comparable results, and the development of manpower in the field of toxicology. Other activities carried out by the IPCS include the development of know-how for coping with chemical accidents, coordination of laboratory testing and epidemiological studies, and promotion of research on the mechanisms of the biological action of chemicals.

WHO Library Cataloguing in Publication Data

Biomarkers and risk assessment : concepts and principles.

(Environmental health criteria ; 155)

1.Biological markers   2.Environmental exposure   3.Hazardous substances
4.Risk factors   I.Series

ISBN 92 4 157155 1       (NLM Classification: QH 541.15.B615)
ISSN 0250-863X

PRINTED IN FINLAND
93/9820 – VAMMALA – 7000

**CONTENTS**

BIOMARKERS AND RISK ASSESSMENT: CONCEPTS AND
PRINCIPLES

# WHO TASK GROUP ON BIOMARKERS AND RISK ASSESSMENT: CONCEPTS AND PRINCIPLES

*Members*

Dr A. Aitio, Institute of Occupational Health, Helsinki, Finland (*Chairman*)[a,b]

Dr D. Anderson, British Industrial Biological Research Association, Carshalton, Surrey, United Kingdom (*Rapporteur*)[a,b]

Dr P. Blain, Division of Environmental and Occupational Medicine, The Medical School, University of Newcastle upon Tyne, United Kingdom[b]

Dr J. Bond, Chemical Industry Institute of Toxicology, Research Triangle Park, North Carolina, USA[b]

Dr M. Buratti, Clinica del Lavoro, Instituti Clinici di Perfezionamento, Milan, Italy[a]

Dr I. Calder, Occupational and Environmental Health, South Australian Health Commission, Adelaide, South Australia, Australia[b]

Dr I. Chahoud, Institute of Toxicology and Embryo-pharmacology, Free University of Berlin, Berlin, Germany[a]

Dr J.R. Fowle, Health Effects Research Laboratory, US Environmental Protection Agency, Research Triangle Park, North Carolina, USA[a]

Dr L. Gerhardsson, Department of Occupational and Environmental Medicine, Lund University Hospital, Lund, Sweden[b]

Dr. R. Henderson, Lovelace Inhalation Toxicology Research Institute, Albuquerque, New Mexico (*Vice-Chairman*)[a,b]

Dr H.B.W.M. Koëter, TNO-CIVO Institute, AJ Zeist, The Netherlands[a]

Dr A. Nishikawa, Division of Pathology, National Institute of Hygienic Sciences, Tokyo, Japan[b]

Dr C.L. Thompson, Laboratory of Biochemical Risk Analysis, National Institute of Environmental Health Sciences, Research Triangle Park, North Carolina, USA[b]

Dr H. Zenick, Health Effects Research Laboratory, US Environmental Protection Agency, Research Triangle Park, North Carolina, USA[b]

*Observers*

Dr J. Lewalter, Institute of Biological Monitoring, Medical Department, Bayer AG, Leverkusen, Germany (attending on behalf of CEC[a] and ECETOC[b])

Dr D. Howe, Unilever E.S.L., United Kingdom (attending on behalf of ECETOC[b])

Mrs G. Richold, Unilever E.S.L., United Kingdom (attending on behalf of ECETOC[b])

*Secretariat*

Dr G.C. Becking, International Programme on Chemical Safety, Interregional Research Unit, World Health Organization, Research Triangle Park, North Carolina, USA[a]

Dr J. Hall-Posner, Unit of Mechanisms of Carcinogenesis, International Agency for Research on Cancer, Lyon, France[b]

Dr F. He, World Health Organization, Division of Health Protection and Promotion, Occupational Health, Geneva, Switzerland[b]

Dr A. Robinson, Ontario Ministry of Labour, Toronto, Ontario, Canada (*Temporary Adviser*)[b]

---

[a] Participant in Planning Meeting on Utilization of Biological Markers in Risk Assessment (Non-Carcinogenic End-Points), 25-27 October 1989, Carshalton, UK

[b] Participant in Task Group on Biomarkers and Risk Assessment: Concepts and Principles, 16-20 November 1992, Carshalton, UK

## NOTE TO READERS OF THE CRITERIA MONOGRAPHS

Every effort has been made to present information in the criteria monographs as accurately as possible without unduly delaying their publication. In the interest of all users of the Environmental Health Criteria monographs, readers are kindly requested to communicate any errors that may have occurred to the Director of the International Programme on Chemical Safety, World Health Organization, Geneva, Switzerland, in order that they may be included in corrigenda.

\* \* \*

A detailed data profile and a legal file can be obtained from the International Register of Potentially Toxic Chemicals, Case postale 356, 1219 Châtelaine, Geneva, Switzerland (Telephone No. 9799111).

\* \* \*

This publication was made possible by grant number 5 U01 ES02617-14 from the National Institute of Environmental Health Sciences, National Institutes of Health, USA.

## BIOMARKERS AND RISK ASSESSMENT: CONCEPTS AND PRINCIPLES

At the Sixth Meeting of the IPCS Programme Advisory Committee (31 October to 3 November 1989) it was recommended that the IPCS give priority to work on biomarkers, as outlined at an IPCS Planning Meeting (25-28 October 1989). One of the recommendations from the Planning Meeting was for the IPCS to prepare an Environmental Health Criteria monograph on the concepts and principles supporting the use of biomarkers in the assessment of human health risks from exposure to chemicals.

The drafts of this monograph were prepared by Dr A. Robinson, Toronto, Canada. During the preparation of the monograph many scientists made constructive suggestions and their assistance is gratefully acknowledged.

A WHO Task Group on Biomarkers and Risk Assessment: Concepts and Principles met in Carshalton, United Kingdom, from 16 to 20 November 1992. Dr Robinson opened the meeting on behalf of the heads of the three cooperating organizations (UNEP/ILO/WHO), and Dr D. Anderson welcomed the participants on behalf of the British Industrial Biological Research Association, the host institution. The Task Group reviewed and revised the draft monograph.

Following the Task Group Meeting, Dr Robinson collated the text with the assistance of Dr A. Aitio and Dr D. Anderson, Chairman and Rapporteur, respectively, of the Task Group. The Secretariat wishes to acknowledge their special contributions in finalizing this monograph.

Dr A. Robinson was responsible for the overall scientific content, and Dr P.G. Jenkins (IPCS Central Unit) for the technical editing.

The efforts of all who helped in the preparation and finalization of the monograph are gratefully acknowledged. Special thanks are due to the United Kingdom Department of Health for its financial support of both the Planning and the Task Group Meetings.

# PREFACE

The purpose of this monograph is to examine the concepts and to identify the principles for the application of biomarkers to assessment of risk to human health from exposure to chemical agents, with special attention to criteria for selection and validation.

Information and examples are provided to illustrate and assist the application of these principles to enhance human health risk assessment by reducing the uncertainties associated with the process. Biomarkers may be indicative of exposure, effect(s) or susceptibility of individuals to chemical agents, but their use must take account also of ethical and social considerations.

Some guidance is provided for the selection of appropriate biomarkers to allow identification of individuals and sub-populations at increased risk, with consequent implications for administrative intervention, mitigation and health protection.

A review has been made of biomarkers suitable for application to assessment of the risk of chemicals that are toxic to the hepatic, renal, haematological, immune, pulmonary, reproductive/developmental and nervous systems or are associated with carcinogenic mechanisms. However, greater detail is provided for biomarkers linked with carcinogenesis, reflecting the volume of scientific publications resulting from recent intensive studies of mechanisms, provoked by public attitudes and perceptions associated with diagnosis of the disease. This section serves to illustrate the complexity of the interactions and the many factors which will influence selection and application of biomarkers to improve further the process of health risk assessment.

# 1. INTRODUCTION

## 1.1 Biomarkers - concepts

Analysis of tissues and body fluids for chemicals, metabolites of chemicals, enzymes and other biochemical substances has been used to document the interaction of chemicals with biological systems. Measurements of these substances, now referred to as "biomarkers", are recognized as providing data linking exposure to a chemical with internal dose and outcome and as relevant to the process of risk assessment.

The term "biomarker" is used in this monograph, as it is in the US National Academy of Sciences report (US NRC, 1989b), in a broad sense to include almost any measurement reflecting an interaction between a biological system and a potential hazard, which may be chemical, physical or biological. The measured response may be functional and physiological, biochemical at the cellular level, or a molecular interaction. Various factors will apply in assessing risks to individuals and population subgroups compared with the general population.

In the assessment of risk, biomarkers may be used in hazard identification, exposure assessment and to associate a response with the probability of a disease outcome. By examining the interactions between human host and chemical exposure, and comparable data for experimental studies of mammalian species, criteria for the selection of biomarkers indicative of exposure, effects, susceptibility and toxic response(s) to chemicals may be established.

The reaction to exposure to a chemical depends on inherited and acquired characteristics and the life-style of the human subject (or other biological system), the properties and form of the chemical, and the circumstances of the contact. The outcome may be no effect, some adverse effect with recovery, or toxicity with morbidity.

Human health is affected by all the activities of an individual, who is subject to a continuum of chemical exposures in the external environment, including air, water, soil and food. It should be noted that distinction of exposure to chemicals on the basis of context, such as recreational, residential or occupational, is often made for administrative convenience. The important considerations for assessment of risk are the dose rate, route, duration and frequency of exposure.

The application of biomarkers, linked to toxic processes or mechanisms, to the risk assessment process, and particularly to quantitative risk assessment, has the potential to provide a more rational and less judgmental process, particularly when compared with methods that arbitrarily attach protection factors to doses assessed to minimize or avoid effects deemed adverse to health.

Selection of appropriate biomarkers is of critical importance because of the opportunity for greater precision in the assessment of risk to individuals or population sub-groups, with the consequent implications for mitigation and health protection. However, selection will depend upon the state of scientific knowledge and be influenced by social, ethical and economic factors.

Subject to ethical considerations, the use of validated biomarkers to monitor exposed populations may provide the basis for early, health-protective intervention.

Identification of practicable biomarkers associated with different toxic end-points or outcomes requires interdisciplinary cooperation and research, and this is evident in relation to carcinogenesis, neurotoxicity, pulmonary toxicity, immunotoxicity and human reproduction. While not all of these areas of interest are equally well developed, use of biomarkers linked with toxicity should enhance the process and reliability of predictions of risk.

Improved definition of the risk associated with exposure to chemicals will permit effective preventive intervention to protect human health both in general and in particular circumstances. Protective measures may include avoidance of exposure to chemicals or protection of sensitive individuals.

## 1.2 Definitions

The term "biomarker" is used in a broad sense to include almost any measurement reflecting an interaction between a biological system and an environmental agent, which may be chemical, physical or biological. However, discussion in this monograph is limited to chemical agents. Three classes of biomarkers are identified:

• *biomarker of exposure*: an exogenous substance or its metabolite or the product of an interaction between a xenobiotic agent and some target molecule or cell that is measured in a compartment within an organism;

• *biomarker of effect*: a measurable biochemical, physiological, behavioural or other alteration within an organism that, depending upon the magnitude, can be recognized as associated with an established or possible health impairment or disease;

• *biomarker of susceptibility* - an indicator of an inherent or acquired ability of an organism to respond to the challenge of exposure to a specific xenobiotic substance.

## 1.3 Biomarkers and the risk assessment process

For a general discussion of concepts and principles underlying assessment of risk to human health associated with exposure to chemicals, the reader is referred to WHO (in press).

The process for assessment of human health risks associated with exposure to chemicals is multifaceted and incorporates the following major components:

• *hazard identification*: to confirm that the chemical is capable, subject to appropriate circumstances, of causing an adverse effect in humans;

• *dose-response assessment*: to establish the quantitative relationship between dose and effect in humans;

• *exposure assessment*: to identify and define the exposures that occur, or are anticipated to occur, in human populations.

Risk characterization is the synthesis of the qualitative and quantitative information that describes the estimated risk to human health from the anticipated environmental exposure.

Hazard identification and dose-response assessment make use of all available data for human and test species and, where appropriate, for *in vitro* test systems.

The relevance of biomarkers to the phases of the risk assessment process is discussed more fully in later sections that address biomarkers of effects, exposure and susceptibility.

# 2. USES OF BIOMARKERS

Biomarkers may be used to assess the exposure (absorbed amount or internal dose) and effect(s) of chemicals and susceptibility of individuals, and they may be applied whether exposure has been from dietary, environmental or occupational sources. Biomarkers may be used to elucidate cause-effect and dose-effect relationships in health risk assessment, in clinical diagnosis and for monitoring purposes.

Biomarkers of exposure can be used to confirm and assess the exposure of individuals or populations to a particular substance, providing a link between external exposures and internal dosimetry. Biomarkers of effect can be used to document either preclinical alterations or adverse health effects elicited by external exposure and absorption of a chemical. Thus the linkage of biomarkers between exposure and effect contributes to the definition of dose-response relationships. Biomarkers of susceptibility help elucidate the degree of the response to exposure elicited in individuals.

## 2.1 Use in health risk assessment

Measurements carried out for many years within the context of "biological monitoring" have been used to assess worker exposure and, in clinical settings, to evaluate the administration of therapeutic agents. These measurements, or biomarkers, provide the critical link between chemical exposure, internal dose and health impairment, and are of value in assessment of risk. However, there is a need to identify and validate for each organ system those characteristic parameter(s) that are indicative of induced dysfunction, clinical toxicity or pathological change, as well as to establish the specificity and sensitivity of each biomarker and its method of measurement.

## 2.2 Use for clinical diagnosis

Biomarkers may be used to:

- confirm diagnosis of acute or chronic poisoning;
- assess the effectiveness of treatment; and
- evaluate the prognosis of individual cases.

For this purpose, a well-established relationship between biomarker(s) and outcome must be available. Assessment of exposure in short-term or long-term exposure situations can be evaluated on a more meaningful basis where previous exposure has been documented by consecutive measurements over a period of time. Although this may not be possible in the circumstances of a major chemical release, biomarkers of effect may still find useful application to assess clinical outcome(s).

## 2.3 Use for monitoring purposes

Biomarkers may be used to confirm the exposure of individuals in a population to a particular substance, e.g., an organic solvent in exhaled breath, the cadmium burden of the kidney, lead in bone, or the fatty tissue storage of chlorinated hydrocarbons (see Table 1, chapter 5). Quantitative measurements may facilitate the determination of dose-response relationships.

Biomarkers are used for screening and for monitoring (repeated at timed intervals), and may be determined and applied on an individual basis or may be related to a population group. Population groups "at risk" may be identified by deviations from normal of mean values for biomarkers of exposure or effects; individual variations will be reflected in statistical terms.

Some public and occupational health surveillance programmes include the use of biomarkers for screening and monitoring purposes. Although the terms "biological screening or monitoring" and "health monitoring" have been applied, there is no agreement that the terms are appropriate, and repeated measurement of biomarkers may be cost-effective methodologies to monitor disease development. In practice, however, ethical and social considerations, rather than cost, often preclude the widespread use of biomarkers for monitoring or surveillance purposes.

Biomarkers of exposure or effect may be used to evaluate compliance with advice for minimizing exposure or for remedial measures in a public health context, e.g., to confirm reduced exposure to lead from environmental sources in a population group. In addition, biomarkers may be used to supplement environmental or ambient workplace measurements of chemicals with recognized potential adverse health effects that may be subject to regulatory controls.

Biomarkers may serve as a basis for assessing individual or population groups exposed to chemicals from any source, including life-style activities. In an occupational context, biomarkers will provide a supplementary means for reviewing the adequacy of protective measures, including work practices and working conditions.

When the inter-individual variation of the biomarker is large in comparison to intra-individual variation, analysis of paired samples (before, during and after the exposure) may greatly enhance the power of the biomarker to detect exposure, e.g., serum acetylcholine esterase measurements in relation to exposure to an organophosphorus compound.

Use of biomarkers reflecting genetically linked or acquired susceptibility to specific chemicals or their metabolites provides an opportunity for the recognition and protection of sensitive individuals. The classic example of genetically linked susceptibility is phenylketonuria in newborn infants. An example of acquired susceptibility is the development of hypersensitivity to certain inhaled gases or dusts (such as toluene diisocyanate, trimellitic acid anhydride or cotton dust) in the workplace.

# 3. SELECTION AND VALIDATION OF BIOMARKERS

The process of selection and validation requires careful consideration of the specificity and sensitivity of the biomarker as a measure of the contribution of the exposure to an observed adverse health outcome. A similar process must be applied also to establishing the accuracy, precision and quality assurance of the analytical procedure for measurement of the selected biomarker.

Before discussing the criteria for the selection and validation of biomarkers of exposure, effect and susceptibility, and their application to the risk assessment process, it is necessary to consider key factors that can influence the host reaction to xenobiotic chemicals. Fig. 1 summarizes some of the various factors that influence the interaction between host and chemical.

These factors may be considered in the context of a source-chemical-host response where the source of the specific chemical of concern may be the air, water, soil or food. It is important to consider the physico-chemical properties of the chemical (e.g., gas, vapour, particle) and whether the chemical is present in a complex chemical mixture or adsorbed on a particle. For example, the initial site of deposition (and perhaps the site of toxicity) of a chemical in the respiratory tract may be affected by the strength of the association between the chemical and particulate matter (which determines bioavailability), as well as by particle size, in the inhaled atmosphere (e.g., nose versus deep lung). Several exposure characteristics need also to be considered, such as the concentration of the chemical and the duration, frequency and magnitude of exposure. Exposure of the host can be through various routes including the respiratory tract (inhalation exposure), the gastrointestinal tract (oral exposure) and the skin (dermal exposure). Finally, there are a number of host characteristics that can influence response to chemical exposure, including age, race, gender, health status, genetic susceptibility, and previous exposure to the same or other chemicals. Information relating to these factors can provide clues as to the types of biomarkers that may be used to assess exposure, effect and susceptibility.

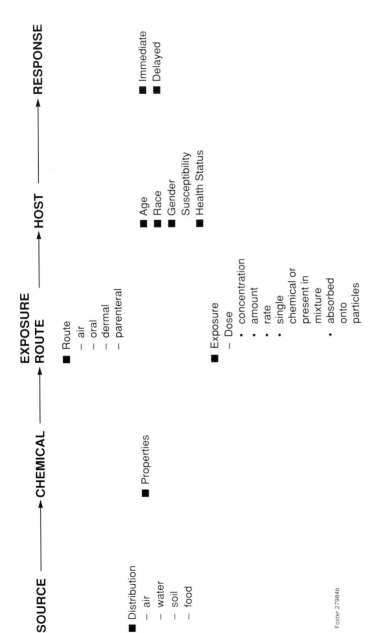

Fig. 1. Some critical factors influencing interaction between host and chemical

Many factors require consideration in the process for selection and validation of a biomarker. To select the most appropriate biomarker requires several steps:

(1)     the identification and definition of the end-point of interest;

(2)     the assembly of the data base to document the relationship between the chemical exposure, the possible biomarkers and the end-point. This will include data from *in vitro*, mammalian and human studies, with assessment of the validity of data and the study protocols;

(3)     selection of biomarker(s) specific to the outcome of interest with careful consideration of the biomarker to identify what is being quantified, to assess the sensitivity and specificity of the marker in relation to exposure, and the significance with respect to health outcome or pathological change over time;

(4)     consideration of specimens potentially available for analysis, with emphasis on protecting the integrity of the specimen between collection and analysis, and a preference for non-invasive techniques;

(5)     review of the analytical procedures available for quantification of biomarkers and their limitations with respect to detection limit, sensitivity, precision and accuracy;

(6)     establishment of an appropriate analytical protocol with provision for quality assurance and quality control;

(7)     evaluation of intra- and inter-individual variation for a non-exposed population;

(8)     analysis of the data base to establish dose-effect and dose-response relationships and their variation, with special emphasis on susceptible individuals;

(9)     calculation or prediction of risk to human health either for the general population or a sub-group; and

(10)    review of ethical and social considerations.

These issues are discussed more fully in the following sections.

These steps may need to be carried out in an interactive and iterative manner before selection of the desired biomarker can be made.

## 3.1 Selection - practical aspects

### 3.1.1 General laboratory considerations

Measurement of biomarkers may range from molecular events to functional outcomes such as behaviour or pulmonary function; the same consideration should be applied to all biomarkers.

Analytical considerations include defined and appropriate precision and accuracy, and quality assurance and control, as well as the availability of automated instrumentation or alternative simple, but specific, methodology. Specimen collection, handling and storage should require the minimum of special precautions to avoid contamination and/or deterioration. The costs, in terms of skilled human resources, equipment and reagents should be reasonable.

Sampling and measurements should preferably be:

- non-invasive;
- representative, i.e. the time of the exposure in relation to measurement should be taken into account; and
- the stability of the analyte in the specimen should be established.

In general terms, specimens available for analysis will include blood, urine, sputum, saliva, finger-nails, breath, hair, faeces and (shed) teeth. The clinic or hospital setting may provide the opportunity to collect unique fluid samples (e.g., follicular, amniotic, semen) or tissues accompanying examination of the patient (e.g., cytological material, pulmonary lavage), tissue biopsies (e.g., fat) or autopsy specimens.

Specialized techniques for *in vivo* determination may be available for some chemicals, e.g., cadmium in kidney or lead in bone, but such applications require exposure of individuals to radiation. In such instances, ethical considerations must also be taken into account.

### 3.1.2 Quality assurance and control

Critical to the successful and effective application of biomarkers is a well-documented quality assurance and control programme. It is beyond the scope of this monograph to discuss such programmes in detail, but these have been reviewed (Aitio, 1981; WHO, 1992b). It is important to note that good analytical performance does not necessarily guarantee accurate results in biomarker analyses since greater errors may be introduced during sampling. Thus, the quality assurance protocol has to cover the entire process. Major impediments to quality assurance of biomarker analyses include the lack of certified reference materials and external quality control programmes.

## 3.2 Validation and characteristics of biomarkers

Validation is a process to establish the qualitative and quantitative relationship of the biomarker (a) to exposure to a chemical, and (b) to the selected end-point. Desirable characteristics of biomarkers include that:

(1) the marker (measurement)
- (a) reflects the interaction (qualitative or quantitative) of the host biological system with the chemical of interest,
- (b) has known and appropriate specificity and sensitivity to the interaction,
- (c) is reproducible qualitatively and quantitatively with respect to time (short- and long-term);

(2) the analytical measurement has defined and appropriate accuracy and precision;

(3) the marker is common to individuals within a population or subgroup and is of defined variability within the normal, non-exposed population or group of interest; and

(4) the marker is common between species.

# 4. ETHICS AND SOCIAL CONSIDERATIONS

It is important to recognize the ethical and social implications of the uses of biomarkers, in addition to the scientific and cost considerations. Ethical concerns may limit the extent of investigations of chemically exposed human individuals and populations, particularly those involving the living.

Participation by individuals or groups will be influenced both by personal and scientific factors. Personal attitudes, ideals and beliefs will vary geographically with ethnic origin and cultural practices.

The process leading to participation is critically important and must respect the dignity, rights and freedom of choice of individuals; participation must be voluntary and based on full information.

The freedom of choice of individuals will include the right of refusal to give blood or other biological samples for analysis of biomarkers. Personal decisions should be based on full information with the implications of a refusal being explained and understood.

Before measurement of a biomarker is undertaken, there should be consideration of how and to whom results should be provided, the interpretation of the results, and whether this should be on an individual or group basis, with or without protection of confidentiality. While practices will vary between countries, it is particularly important that the role of medical officers be defined in relation to their responsibility to the individual (patient) and in relation to administrative (company) management.

It is the ethical responsibility of medical officers to inform individuals fully of potentially hazardous exposures, recognizing that remedial action may involve administrative decisions. The latter may be taken in the context of prevailing economic and social considerations rather than of individual circumstances.

Investigators need to recognize and accommodate the medical dilemma created by the conduct of biomarker research in terminally ill patients, since the data may clarify the use of the biomarker without contributing to improvement of the health status of the patient. Thus, careful consideration must be given to the desire to advance scientific understanding relative to meeting the needs of the patient.

In addition to the general ethical issues associated with use of biomarkers, there are particular problems relating to biomarkers of susceptibility.

Identifying susceptible individuals may help to prevent their exposure to a specific harmful chemical(s) but may lead to discrimination in the employment of the susceptible individual when that chemical is known to be present in the workplace.

It is also an ethical question as to whether individuals should be given information about their own susceptibility. This knowledge would allow them to make more informed choices, for example, the avoidance of exposure to specific substances. It is, however, important to realize that only for a few biomarkers of susceptibility is it well established that they are associated with the development of disease. If the individual does not fully understand this uncertainty, such information on biomarkers may cause unnecessary concern and anxiety.

# 5. BIOMARKERS OF EXPOSURE

The exposure assessment component of the risk assessment process is an attempt to provide qualitative and quantitative estimates of human exposure through the use of measurements and models. In this context, measurements may be made of chemical concentrations in food, water and air, selected environmental concentrations (e.g., occupational or residential settings) as well as measures of the actual exposures experienced by the individual or population. Exposure biomarkers extend this latter component of exposure assessment into the realm of data which provide the most direct evidence of human exposure to a given agent and the absorbed dose.

Adverse or toxic effects in a biological system are not produced by chemical agents unless that agent or its biotransformation products reach appropriate sites in the body at a concentration and for a length of time sufficient to produce the toxic manifestation. Thus, to characterize fully the potential hazard or toxicity of a specific chemical agent in an individual, it is necessary to identify not only the type of effect and the dose required to produce the effect but also information about the duration and frequency of exposure to the agent, and the susceptibility of the exposed individual.

Methods for assessing exposure to a chemical fall into two categories:

1. measurement of levels of chemical agents and their metabolites and/or derivatives in cells, tissue, body fluids or excreta; and

2. measurement of biological responses such as cytogenetic and reversible physiological changes in the exposed individuals.

Measurement of covalent adducts formed between chemical agents and cellular macromolecules (proteins, DNA), or their excretion products have characteristics of both category one and two above.

In evaluating exposure, distinction is made between the *external dose*, defined as the amount of a chemical agent in environmental contact with the organism, as determined by personal or area monitoring, and the *internal dose*, which is the total amount of a chemical agent absorbed by the organism over a period of time. Biomarkers of exposure will reflect the distribution of the chemical or its metabolite throughout the organism. Theoretically, this distribution can be tracked through various biological levels (e.g., tissue, cell, etc.) to the ultimate target. The concept of biomarkers of exposure is illustrated in Fig. 2.

This figure illustrates that, of the total amount absorbed, only a portion will be delivered to a target tissue. A portion will reach internal macromolecules, and a smaller proportion will reach the critical site on the macromolecule, with only a fraction of the latter amount acting as the biologically effective dose. Biomarkers for each of these forms of internal dose would be useful for assessment of risk. The decreasing area of the triangle, from the total absorbed amount to each lower level as distribution and metabolism occur, illustrates the decreasing mass of the internal dose that reaches the target tissue, cell, or critical site. In progressing from biomarkers of total absorbed dose to markers of biologically effective dose, it becomes increasingly easy to relate the dose to the mechanism of the induced health effect. In the reverse process, from the dose for critical sites to the total amount absorbed, it becomes easier to relate the internal dose to the external exposure.

The internal dose can be assessed by suitable analyses of biomarkers in biological samples (urine, faeces, blood and/or its components, expired air) (Alessio et al., 1983, 1984, 1986, 1987, 1988, 1989; UK HSE, 1991; ACGIH, 1992; DFG, 1992; Bond et al., 1992). These biomarkers may be the unchanged chemical material, its known metabolites or biochemical markers affected by absorption of the chemical. It may be possible to estimate the dose quantitatively when the toxicokinetics of the chemical is well established and the sampling is conducted at appropriate points in time. The tissue dose may be further refined to a specific target dose which may be defined as the amount of chemical (or its metabolite) which, over a period of time, reaches the biologically significant site(s) within the target tissue.

While such measurements may not equate to the biologically effective dose, the data can provide useful estimates of internal exposure (dose). Knowledge of the kinetics of formation and removal from the body of these types of biomarkers provides a link between exposure and internal dose.

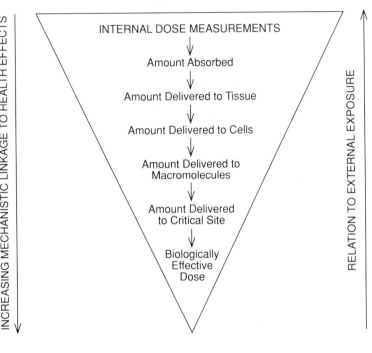

Fig. 2. Biomarkers for internal dose for chemicals for which the major mechanism of action occurs through molecular interaction

Specific measures of internal dose are the active chemical species (either parent compound or metabolite) delivered to target tissues or cells, the reactive chemical species delivered to target organelles or macromolecules, or the reactive chemical species that participates in biochemical reactions. For example, quantification of the generalized covalent binding of reactive species to macromolecules will provide a measure of absorbed dose delivered to target tissues or cells, while measurement of total DNA adducts is indicative of the dose delivered to target organelles or macromolecules. Finally, specific DNA adducts could be the biologically effective species that initiate the carcinogenic process. These issues are discussed in chapter 7.

The potential impact of target tissue DNA-protein cross-links as a biomarker of the biologically effective dose is illustrated in the proposed risk assessment for formaldehyde put forward by US EPA (1991). The biological, mechanism-based model of formaldehyde carcinogenesis consists of three submodels (Conolly et al., 1992; Conolly & Andersen, in press). One of these is a tissue dosimetry submodel which incorporates formaldehyde-induced cross-links of DNA with adjacent proteins (Casanova et al., 1989, 1991). Although the role of DNA-protein cross-links in the mechanism of formaldehyde-induced nasal cancer is not known, their formation is used only as a biomarker of the "biologically effective" dose reaching target cells in the nasal cavity. An earlier proposal (US EPA, 1987) used external formaldehyde concentration as the measure of dose. The predicted quantitative human risk at low levels of exposure is lower when the interspecies extrapolation (from monkeys and rats) is based on the biomarker, i.e. DNA-protein cross-links, rather than on the concentration of formaldehyde inhaled. Although, there are several unresolved issues regarding the use of DNA-protein cross-links as a biomarker, this example illustrates that mechanistic data and a biomarker of delivered dose can be used in the risk assessment process for chemicals.

A strategy to help relate biomarkers to prior exposures is to obtain quantitative information about the kinetics of formation and breakdown of the biomarker, as shown in Fig. 3. The information required includes the quantified correlation of the biomarker with a given exposure scenario and kinetic data for elimination of the biomarker. Biomarkers for chemicals that are cleared rapidly, such as vapours in exhaled breath or urinary metabolites, may be present in large amounts soon, during or

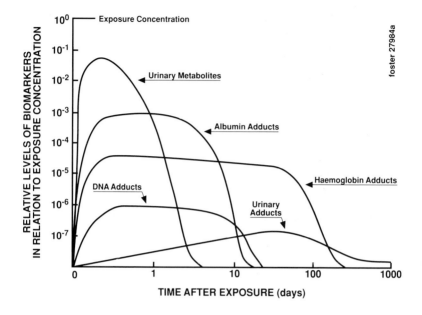

Fig. 3. Hypothetical relationships among different biomarkers of exposure
with respect to their relative levels and time of appearance after a single dose
(Henderson et al., 1989)

immediately following an exposure, but are not detectable at later times.

Other biomarkers, such as adducts formed with blood proteins, may represent only a small fraction of the total internal dose but, because they have a long half-life in the body (relative to exposure frequency), may accumulate to detectable levels with continued exposure.

Use can be made of kinetic properties to define prior exposures. If a person has had only a single, recent exposure to a chemical, the level of biomarkers with short half-life will be high relative to those with a longer half-life. With continuing exposure, the levels of markers with both shorter and longer half-lives should be high. If a person was exposed in the more distant, rather than the more recent, past, only the biomarkers with the longer half-lives will be detectable.

Thus by analysing several biomarkers with different half-lives (e.g., haemoglobin adducts in the blood, metabolites of the chemical in urine, parent compound in blood) from a single individual at a single point in time, more information may be obtained about the nature of the past exposure than from use of a single biomarker.

Also of value in the interpretation of biomarker data is the use of mathematical models to describe the kinetics of formation and elimination of biomarkers of exposure. Absorbed chemicals are distributed between various compartments in the body, with the distribution being dependent on the nature of the compartment and the lipophilicity of the chemical. The most simple models use only one compartment; however, multicompartmental models are usually required to describe the disposition of most chemicals in the body (Gibaldi & Perrier, 1982). Multicompartmental models that incorporate biomarkers of exposure in relation to toxic end-points are well established. A biokinetics model for lead has been used to predict blood lead levels of individuals and communities (DeRosa et al., 1991).

More recently, physiologically based toxicokinetic (PBTK) models have been developed that make use of the physico-chemical properties of a chemical (such as partition coefficients that indicate how the chemical or its metabolites become partitioned between different fluids in the body), the kinetics of metabolism of the chemical (such as $V_{max}$ and $K_M$ for metabolic pathways), and physiological parameters of the exposed individual (such as tissue blood flow, respiratory minute volume, cardiac output) to predict the actual concentrations of biomarkers that will occur after specific exposure regimens. These models are adapted for humans by changing the physiological and metabolic parameters to those appropriate for humans, and by testing for validity in limited human studies. These models can then be used to extrapolate between different exposure situations for the predicted levels of markers (Ramsay & Andersen, 1984; Droz et al., 1989).

Another use of models is in defining the quantitative relationship between biomarkers in readily available biological samples (e.g., blood cells) and in those less readily available which might be more pertinent biomarkers for the health effect of concern (e.g., tissue DNA). An example is the use of haemoglobin adducts to predict the amount of DNA adducts at a critical site, e.g., after exposure to ethylene oxide (Passingham et al., 1988). To be able to make such predictions, one must know the kinetics of formation and breakdown of each of the markers and the factors that influence those kinetics. Based on such information a model can be developed to show the quantitative relationships between the markers under different exposure conditions or at different times after exposure.

Biomarkers are used extensively in the surveillance of workers occupationally exposed to metals such as lead, cadmium, mercury, nickel, chromium and arsenic, and to organic chemicals such as aniline, benzene, carbon disulfide, styrene, chlorobenzene and chlorinated aliphatic hydrocarbon solvents (see Table 1). The examples in Table 1 are given as general information. Before applying them in specific circumstances, readers must consult the original references.

These measures are used to indicate the absorbed dose. For a few chemicals, notably lead, mercury, cadmium and carbon monoxide, an approximate estimation of the associated health risk may be made. For other chemicals, exposure may be assessed in quantitative terms. It is noted that biomarkers are supplementary to environmental measurements rather than alternative or substitute measures of exposure.

Considerable efforts are being made to develop biomarkers associated with exposure to chemical carcinogens and to establish the relationship between a marker and the future health risk. The use of animal models may facilitate this process; studies on the DNA adducts formed by vinyl chloride illustrate the types of strategies required to make each link (Swenberg et al., 1990). In rats, vinyl chloride induces liver tumours with a high incidence in young animals, and the DNA adducts (biomarkers) formed in the liver have been characterized and the half-lives determined. DNA fidelity replication assays were used to show one type of adduct that had both a long half-life and was capable of inducing mutations (Hall et al., 1981; Barbin et al., 1985; Singer et al., 1987; Swenberg et al., 1990). The level of this adduct was much higher in the livers of young rats exposed to vinyl chloride than in adults, and was characterized as the one most closely related to the health effect. To extend this animal research to predict human health risks, it would be necessary to determine if the concentration of the same adduct is associated with liver tumour formation in human tissue samples.

Table 1. Some parameters proposed for biological monitoring by different organizations

| Exposure | Measured parameter[a] | | | |
|---|---|---|---|---|
| | American Conference of Governmental Industrial Hygienists (ACGIH, 1992) | Deutsche Forschungs-gemeinschaft (DFG, 1992) | Finnish Institute of Occupational Health (FIOH, 1993) | United Kingdom Health and Safety Executive (UK HSE, 1991) |
| Acetylcholinesterase inhibitors | E-acetylcholinesterase | E-acetylcholinesterase | E-acetylcholinesterase | E-acetylcholinesterase, P-cholinesterase |
| Aluminium (Al) | | U-Al | U-Al | |
| Aniline | U-p-aminophenol,B-Met-Hb | U-aniline | U-p-aminophenol | |
| Arsenic (As) | U-certain volatile arsenic compounds produced by direct hydrogenation | U-certain volatile arsenic compounds produced by direct hydrogenation | U-As$^{III}$+, As$^{V}$+ | U-As$^{III}$+, As$^{V}$+ MMA+DMA |
| Benzene | U-phenol | B-benzene, U-phenol | B-benzene | Breath-, B-benzene |
| p-tert-Butylphenol | | U-p-tert-butylphenol | | |
| Cadmium (Cd) | U-Cd, B-Cd | U-Cd, B-Cd | U-Cd, B-Cd | U-Cd, B-Cd |
| Carbon disulfide | U-TTCA | U-TTCA | U-TTCA | U-TTCA |
| Carbon monoxide | Breath-CO, B-COHb | B-COHb | B-COHb | B-COHb |
| Chlorobenzene | U-4-chlorocatechol | U-4-chlorocatechol | | |
| Chlorophenols | | | U-tri+tetra-pentachlorophenols | |

Table 1 (contd).

| | | | | |
|---|---|---|---|---|
| Chlorophenoxy acids | | | U-2,4-D + Dichloroprop + MCPA + Mecoprop | |
| Chromium (Cr) | U-Cr | | U-Cr | B-Cr, U-Cr |
| Cobalt (Co) | | | U-Co | U-Co |
| Dichloromethane | | B-COHb, B-dichloromethane | B-COHb | B-COHb, B-dichloromethane |
| Dieldrin | | | P-dieldrin | B-dieldrin |
| Dimethylformamide | U-methylformamide | | U-methylformamide | U-methylformamide |
| Ethylbenzene | U-mandelic acid | | | U-mandelic acid |
| 2-Ethoxyethanol | | U-ethoxyacetic acid | U-ethoxyacetic acid | |
| 2-Ethoxyethyl acetate | | U-ethoxyacetic acid | U-ethoxyacetic acid | |
| Ethylene oxide | | Breath[b], B-ethylene oxide | | |
| Fluoride (F) | U-F | U-F | U-F | U-F |
| Furfural | U-furoic acid | | | U-furoic acid |
| Halothane | | U-trifluoroacetic acid | | |
| Hexachorobenzene | | P/S-hexachlorobenzene | | |

Table 1 (contd).

| Exposure | Measured parameter[a] | | | |
|---|---|---|---|---|
| | American Conference of Governmental Industrial Hygienists (ACGIH, 1992) | Deutsche Forschungs-gemeinschaft (DFG, 1992) | Finnish Institute of Occupational Health (FIOH, 1993) | United Kingdom Health and Safety Executive (UK HSE, 1991) |
| *n*-Hexane | U-hexanedione, Breath-*n*-hexane[c] | U-hexanedione + dihydroxyhexanone | U-hexanedione | |
| Hydrazine | | P-, U-hydrazine | | |
| Lead (Pb) | B-Pb, U-Pb, B-ZPP | B-Pb, U-ALA | B-Pb, B-ZPP | B-Pb, U-ALA, B-ZPP |
| Lindane | | B(P,S)-lindane | B-lindane | B-lindane |
| Manganese (Mn) | | | U-Mn | B-, U-Mn |
| Mercury (Hg) | | B-, U-Hg | B-, U-Hg | B-, U-Hg |
| Methanol | U-methanol, U-formate | U-methanol | U-formate | |
| Methylbromide | | | | B-Br |
| Methyl butyl ketone | | U-hexanedione + dihydroxyhexanone | | |
| Methylene bis(2-chloroaniline) (MOCA) | | | | U-MOCA |
| Methylene dianiline (MDA) | | | | U-MDA |

Table 1 (contd).

| | | | | |
|---|---|---|---|---|
| Methyl ethyl ketone (MEK) | | U-MEK | U-MEK | |
| 2-Methoxyethanol | | | | U-methoxyacetic acid |
| 2-Methoxyethyl acetate | | | | U-methoxyacetic acid |
| Nickel (Ni) | | U-Ni | U-Ni | U-Ni |
| Nitrobenzene | U-p-nitrophenol, B-MetHb | | | |
| Parathion | U-p-nitrophenol, E-cholinesterase | U-p-nitrophenol, E-cholinesterase | | |
| Pentachlorophenol (PCP) | U-PCP, P-PCP | U-PCP, P-PCP | | U-PCP |
| Phenol | U-phenol | U-phenol | | |
| Polychlorinated biphenyls (PCB) | | | S-PCB | B-PCB |
| 2-Propanol | | U-acetone, B-acetone | | |
| Selenium (Se) | | | U-Se | |
| Styrene | U-mandelic acid, U-PGA | U-mandelic acid, U-mandelic acid + U-PGA | U-mandelic acid + PGA | U-mandelic acid |
| Thallium (Tl) | | | | U-Tl |
| Tetrachloroethylene | Breath-, B-tetrachloro-ethylene | Breath-[b], B-tetrachloro-ethylene | | B-tetrachloroethylene |

Table 1 (contd).

| Exposure | Measured parameter[a] | | | |
|---|---|---|---|---|
| | American Conference of Governmental Industrial Hygienists (ACGIH, 1992) | Deutsche Forschungs-gemeinschaft (DFG, 1992) | Finnish Institute of Occupational Health (FIOH, 1993) | United Kingdom Health and Safety Executive (UK HSE, 1991) |
| Tetrachloromethane | | Breath-[b], B-tetrachloromethane | | |
| Tin (Sn) | | | | U-Sn |
| Toluene | U-hippuric acid | B-toluene | B-toluene | B-toluene |
| 1,1,1-Trichloroethane | Breath-1,1,1-trichloroethane, B-trichloroethanol | Breath-[b], B-1,1,1-trichloroethane | B-1,1,1-trichloroethane | B-1,1,1-trichloroethane |
| Trichloroethylene | U-TCA, B-trichloroethanol, U-TCA + trichloroethanol | U-TCA, B-trichloroethanol | U-TCA, U-Trichloroethanol | U-TCA |
| Vanadium (V) | | | U-V | U-V |
| Vinyl chloride | | U-thiodiglycolic acid | | |
| Xylenes | U-methylhippuric acids | U-methylhippuric acids, B- | U-methylhippuric acids | U-methylhippuric acids |

a   The following abbreviations have been used:  E = erythrocyte;  P = plasma;  S = serum;  U = urine;  B = blood;  ZPP = erythrocyte zinc protoporphyrin;  ALA = δ-aminolevulinic acid;  PGA = phenylglyoxylic acid;  MMA = monomethylarsinic acid;  DMA = dimethylarsinic acid;  TCA = trichloroacetic acid;  MCPA = chloromethylphenoxyacetic acid;  2,4-D = 2,4 dichlorophenoxyacetic acid;  TTCA = 4-thio-4-thiazolidine carboxylic acid.

b   DFG (1992) refers to alveolar air

c   ACGIH (1992) refers to end-exhaled air

Note:Readers must consult original references before applying the above examples to specific situations

# 6. BIOMARKERS OF EFFECT

This section focuses on those human biomarkers that can be applied currently or will be in the near future. Biomarkers of effect may be used directly in hazard identification and dose-response assessment components of the risk assessment process. In hazard identification, biomarkers may facilitate screening and/or identification of a toxic agent and characterization of the associated toxicity. Biomarkers that are implicated in toxic mechanism(s) are preferred for quantitative dose-response assessments when extrapolating from existing data to a human situation of concern (e.g., from high to low dose or from test species to humans).

There are wide inter-individual variations in the response to equivalent doses of chemicals. While the outcome of a chemical insult in an individual may be predicted more accurately from biomarkers of effect(s), such biomarkers may not be specific for a single causative agent. Many biomarkers of effect are used in everyday practice to assist in clinical diagnosis, but for preventive purposes an ideal biomarker of effect is one that measures change that is still reversible. Nevertheless, certain biomarkers of nonreversible effects may still be very useful in epidemiological studies or provide the opportunity for early clinical intervention.

A limited range of tissues is available for routine analysis of biomarkers. The more accessible tissues are therefore used as surrogates for the known or putative target tissues. In some instances biomarkers of effect are not mechanistically related to chemically induced lesions, but may represent concomitant, independent changes. Therefore, although an effect (e.g., sister chromatid exchange) is being analysed, the use is conceptually close to assessment of exposure.

## 6.1 Haematological biomarkers

Inhibition of the enzymes in the haem synthesis pathway (e.g., ferrochelatase, levulinate dehydratase) has been used as a marker of effect of exposure to lead. This effect is reflected also in the levels of free erythrocyte protoporphyrin (FEP) and $\delta$-aminolevulinate in the urine. Elevated levels of urinary $\delta$-aminolevulinate are observed at higher lead exposures than changes in FEP, for example, while basophilic stippling of erythrocytes is an even less sensitive biomarker for the effects of lead. However, the effects on haem synthesis are not specific to lead as a causative agent; iron deficiency has a similar effect on FEP. The relationship of these effect biomarkers to toxicity requires further elucidation.

Routine leucocyte, erythrocyte and thrombocyte counts have been used in the surveillance of patients treated with cytostatic drugs and in the monitoring of benzene-exposed workers. The predictive power in relation to benzene-induced aplastic anaemia or leukaemia is limited (Townsend et. al., 1978; Hancock et al., 1984; Lamm et al., 1989). Ferrokinetic measurements, such as plasma iron disappearance half-time, erythrocyte utilization of iron, plasma iron transport rate, or erythrocyte iron turnover rate, have been suggested as biomarkers of myelotoxicity (Rajamaki, 1984).

## 6.2 Nephrotoxicity biomarkers

Several different types of measures have been tested and used as biomarkers of renal damage. These have been classified as functional markers (e.g., serum creatinine and $\beta_2$-microglobulin), urinary proteins of low or high molecular weight (e.g., albumin, transferrin, retinol-binding globulin, rheumatoid factor, immunoglobulin G), cytotoxicity markers (tubular antigens, e.g., BB50, BBA, HF5), enzymes (e.g., N-acetylglucosaminidase, $\beta$-galactosidase) in urine, and biochemical markers (eicosanoids, e.g., 6-keto PGF2α, PGE2, PGF2α and TXB2, fibronectin, kallikrein activity, sialic acid and glycosaminoglycans in urine, and red blood cell negative charges) (Cardenas et al., 1993a,b; Roels et al., 1993).

Biomarkers for nephrotoxicity were reviewed in WHO (1991) and are well validated in relation to exposure to cadmium (WHO, 1992a; Roels et al., 1993) but not in relation to exposure to mercury or lead (Cardenas et al., 1993a,b).

## 6.3 Liver toxicity biomarkers

Effects of chemicals on the liver have been estimated traditionally by measuring the activities of, for example, aminotransferase (most often aspartate or alanine aminotransferase) in the serum, where they are found when liver cells have been damaged and have leaked their contents. Many other enzymes have also been analysed for this purpose; they include 5-nucleotidase, alcohol dehydrogenase, lactate dehydrogenase, isocitrate dehydrogenase, leucine aminopeptidase, glutathione *S*-transferase, ornithine carbamoyl transferase). Since tissues other than liver also contain these enzymes, their activities may be elevated in serum not only after liver damage but also when non-hepatic tissues have been damaged. To overcome this lack of specificity, analysis of specific isoenzymes has been used. Serum activities of enzymes such as alkaline phosphatase and $\gamma$-glutamyl transpeptidase may be used as biomarkers of hepatic damage mainly involving biliary excretion. Several liver function tests can also be used as biomarkers of effects; these include the concentrations of serum proteins synthesized in the liver, e.g., albumin and clotting factors, or serum concentrations of bile acids, also synthesized in the liver, as well as tests for the hepatic excretory function such as bromsulfphthalein half-time. These parameters lack specificity since hepatic viral infections, alcohol and drug use affect these enzymes. Indirect measures of chemically induced change(s) in the cytochrome P-450 enzyme system, using provocation tests, have been proposed as sensitive indicators. However, the relationship to liver damage and disease is not established and the requirement for the administration of a drug limits the use of such tests.

Hepatotoxicity is caused by a number of chemicals that are metabolized by the cytochrome P-450-dependent mixed-function oxidase system to reactive intermediates. For example, carbon tetrachloride has been studied extensively; it is metabolized to a reactive intermediate which initially depletes intracellular glutathione to a level that is no longer protective when the metabolite reacts with critical macromolecules leading to cell death and hepatoxicity. In this example, biomarkers of effect could include glutathione levels, lipid peroxidation or the number of necrotic cells.

## 6.4 Biomarkers of immunotoxicity

The immune system protects the organism against infectious microorganisms and the growth of at least some neoplasms. Reactions of the immune system are influenced by genetic factors, age, nutrition, life-style and health status. Xenobiotics may stimulate or suppress the immune system. After initial sensitization, even a minimal new exposure may lead to an anaphylactic reaction. The immune system may be more sensitive to chemical challenge than any other body system.

Hypersensitivity reactions following inhalation exposure include asthma, rhinitis, pneumonitis and granulomatous pulmonary reactions (see section 6.5). Hypersensitive dermal reactions induced by chemicals include a wide variety of acute, subchronic and chronic changes. Patch testing has been used traditionally as a biomarker for identification of the causative agent of an allergic skin reaction. However, the possibility of inducing hypersensitivity by patch testing has been well documented and should not be overlooked (Adams & Fisher, 1990).

Elevated levels of specific antibodies, usually of the IgE type, may indicate existing sensitization. However, not all individuals with elevated levels are symptomatic and not all symptomatic individuals exhibit elevated IgE levels (Horak, 1985; Sub-Committee on Skin Tests of the European Academy of Allergology and Clinical Immunology, 1989; Nielsen et al., 1992).

Suppression of the immune system increases susceptibility to infections and neoplasia. Changes in the relative abundance of different lymphocyte subpopulations (suppressor and helper T-cells) have been used as biomarkers for the immune suppression (Jennings et al., 1988; Sullivan, 1989; Holsapple, et al., 1991). Individuals with asbestos-induced pleural or pulmonary changes, or asbestos-induced cancer, as well as those heavily exposed to asbestos but without apparent disease, have been reported to exhibit an altered immunological status (e.g., reductions in T-lymphocyte subsets) (Bekes et al., 1987).

In view of the growing incidence of hypersensitivity reactions to chemicals, development and application of biomarkers for immunotoxic effects is important. However, this is made difficult by the current limited understanding of basic immunological mechanisms and the effects of chemicals thereon (US NRC 1992).

## 6.5 Biomarkers of pulmonary toxicity

The most frequently used markers of pulmonary toxicity measure gross effects on pulmonary function (e.g., peak expiratory flow, forced expiratory volume, transfer factors) rather than effects on cells or biochemical processes (US NRC, 1989a). These measures tend to be nonspecific with respect to the causative agent and may overlook effects specific to a certain cell type. Peak expiratory flow measurements can be performed by the exposed individuals themselves at the workplace, at home or elsewhere, and they provide information on the underlying causes of air-way obstruction, allowing a closer association between exposure, atmosphere and response.

Air-way hyperactivity can be assessed by challenge tests using inhalation exposure. Although such tests may assist in identifying the factors causing hypersensitive pulmonary reactions, there is a clear risk of acute reactions, and testing should be carried out by qualified personnel in carefully controlled environments.

Recently, analysis of bronchoalveolar lavage fluid (BALF) has been used to detect lung injury or to follow the progress of pulmonary disease or the efficacy of therapeutic treatment (Reynolds, 1987; Henderson, 1988).

The use of cellular elements as markers of pulmonary disease state has been emphasized in human BALF analysis (Reynolds, 1987).

Total cell counts and differential counts, including use of monoclonal antibody staining to distinguish T-cell subtypes, are used to detect alveolitis and to aid in the diagnosis of interstitial lung disease. High percentages of lymphocytes are indicators of granulomatous processes, such as sarcoidosis, or hypersensitivity pneumonitis. High percentages of neutrophils with some eosinophils indicate possible idiopathic pulmonary fibrosis. Other extracellular components, such as cytokines and other mediators of inflammation, have been used on an experimental basis to answer specific research questions.

The analysis of BALF has been used to define the dose-response characteristics of inhaled or instilled toxins in animal toxicity studies (Henderson, 1988). The most sensitive biomarker of an inflammatory response in the bronchoalveolar region is the number of neutrophils in BALF. The levels of protein and of extracellular enzymatic activity are also useful markers of pulmonary toxicity. Increases in protein concentrations in BALF indicate increased permeability of the alveolar/capillary barrier. Lactate dehydrogenase (LDH) is a cytoplasmic enzyme that is found extracellularly only in the presence of lysed or damaged cells. Beta-glucuronidase or similar lysosomal hydrolytic enzymes are excellent markers for the toxicity of inhaled particles. These particles are phagocytosed by macrophages, and the enzymes are released from activated or lysed macrophages.

The secretion of cytokines from pulmonary macrophages obtained by bronchoalveolar lavage provides markers of developing fibrosis. Recent studies by Piguet et al. (1990) demonstrate that the level of secretion of tumour necrosis factor (TNF) by pulmonary macrophages is associated with quartz-induced fibrotic processes. Lassalle et al. (1990) found elevated secretion of TNF by macrophages obtained from individuals with coal-workers pneumoconioses compared with macrophages from controls. The secretion of platelet-derived growth factor from pulmonary macrophage was elevated in patients with idiopathic pulmonary fibrosis (IPF).

Glutathione (GSH), a tripeptide protective against oxidative stress, is present in BALF, and a decrease in GSH in BALF is a potential marker for oxidative stress. Decreased levels of GSH have been observed in patient with IPF (Cantin et al., 1989) and in animals exposed chronically to diesel exhaust, resulting in pulmonary fibrosis (Henderson, 1988).

Nasal lavage fluid (NLF) also provides markers of response to inhaled toxins. The work of Graham et al. (1988) demonstrates the potential use of NLF analysis to document the influx of neutrophils into the nasal cavity in humans in response to inhaled ozone.

Biomarkers in blood related to lung injury have not been validated. However, the work of Cavalleri et al. (1991) indicates that serum aminoterminal propeptide of type III procollagen (PIIINP) may become useful as an early marker for developing fibrosis. A dose-dependant increase in serum PIIINP was found in individuals exposed to low or high levels of asbestos.

Finally, urinary levels of amino acids associated with the connective tissue of the lung (hydroxyproline, hydroxylysine, desmosine and isodesmosine) have been used as markers of lung injury (Harel et al., 1980; Yanagisawa et al., 1986; Stone et al., 1991). However, such assays are not specific for lung injury and merely indicate the breakdown of connective tissue in any organ in the body.

## 6.6  Biomarkers of reproductive and developmental toxicity

Markers associated with an adverse outcome in reproduction may reflect toxic effects in the male or female or be associated with development during the embryonic, fetal, perinatal or neonatal period (US NRC, 1989b; Mattison, 1991).

Biomarkers for the male reproductive system may include physiological indicators of impaired testicular function, or sperm number or characteristics (including cytogenetics). Measures of hormonal status (i.e. FSH, LH and testosterone) can also be readily obtained from blood and, in the case of testosterone, from urine and saliva. However, these levels are greatly influenced by circadian rhythms and demonstrate large inter- and intra-individual variability. A clearer picture of hormonal status can be obtained by administering GnRH or LH and examining the hormone response to these challenges. Biomarkers for the male reproductive system are rather easily accessible and some even reasonably well validated; such markers are less well developed for the female reproductive system.

Biomarkers indicative of developmental toxicity should also be considered. As is the case for many biological markers of effects, it is often difficult to identify the causative agent in the absence of any specific exposure history. Biomarkers could include measurements of detrimental effects produced by chemical or other exposures during embryonic or fetal stages of development. Irreversible lesions can be embryolethal or result in functional anomalies in the offspring. Examples of biomarkers of developmental toxicity include low birth weight, chromosome anomalies, delayed growth of specific organ systems, mental retardation, and subtle behavioural changes. The changes associated with $F_1$ male-mediated abnormalities have been discussed by Anderson (1990). Some of these biomarkers of developmental effects (malformations, mental retardation) are not biomarkers of effect as far as the individual is concerned, but rather represent the adverse health outcome itself. However, from the point of view of the exposed population, they may be considered as biomarkers since they show that within a population a harmful exposure has taken place.

Several biomarkers have been proposed for use during pregnancy, e.g., early pregnancy loss and assays for genetic defects of the conceptus. The latter comprise both classical cytogenetic studies, as well as specific DNA probes (US NRC, 1989b). The use of urinary human chorionic gonadotrophin (HCG) has been well documented as a biomarker for early fetal loss (US NRC, 1989b). Many different biomarkers have been used to follow the development of the pregnancy and the well-being of the conceptus, but they have not yet been applied to studies of effects of chemicals on pregnancy.

## 6.7 Biomarkers of neurotoxicity

The functions of the nervous system are complex and biomarkers may range from effects of chemicals on neural cellular and molecular processes to neurophysiological and neuro-behavioural measurements of complex functional entities.

Inhibition of plasma and erythrocyte acetylcholine esterase (AchE) provides biomarkers of exposure to organophosphorus compounds and other cholinesterase inhibitors. While erythrocyte cholinesterase is similar to brain cholinesterase, and is therefore an effect biomarker, plasma nonspecific pseudocholinesterase only reflects exposure and is not a marker of CNS effects.

Measures of the function of the peripheral nervous system (e.g., electroneuromyography, nerve conduction velocities, vibration sensitivity) are well defined. Assessment of peripheral nervous system dysfunction associated with exposure to chemicals can be carried out using electroneuromyography at the preclinical stage (Seppalainen et al., 1979).

Some well-established neurophysiological (e.g., evoked potentials, electroencephalography) and neurobehavioural (e.g., the WHO Neurobehavioural Core Test Battery, Cassito et al., 1990) measures may be used as biomarkers to evaluate CNS dysfunction induced by neurotoxicants. These tests must be carried out under well-controlled conditions.

Methods for assessing changes in higher cognitive function (e.g., learning and memory) have been used extensively, e.g., in workers exposed to solvents or heavy metals, but require further refinement.

Available neuroimaging procedures, e.g., computed axial tomography (CAT), magnetic resonance imaging (MRI), nuclear magnetic resonance spectroscopy (MRS) and positron-emission tomography (PET), are considered non-invasive, but some of them require exposure to ionizing radiation. CAT and MRI can be carried out with current clinical techniques to assess chemically induced changes in the brain. The use of MRS and PET can provide a more detailed evaluation of the biochemical status (e.g., rate of energy generation, blood flow, L-glucose metabolism) in the central nervous system. They can be used as biomarkers for assessing exposure to neurotoxicants inducing brain alterations. However, they are expensive, of little use in assessing spinal cord, nerve and muscle changes, and there is only minimal data validating their use in neurotoxicology (US NRC, 1992).

Other promising biomarkers for neurotoxicity in animal studies include glial fibrillary acidic protein (localized in the astrocytes), which increases in localized areas within the brain where injury due to toxicants occurs (O'Callaghan, 1991).

# 7. BIOMARKERS AND CHEMICAL CARCINOGENESIS

As new information about the multistep process of carcinogenesis unfolds, it is instructive to consider the various mechanisms by which chemicals induce cancer. Knowledge of mechanisms will enable the selection of appropriate biomarkers for use in risk assessment of carcinogens. Some chemicals are direct acting and others require metabolic activation. Once absorbed most chemicals undergo enzyme-mediated reactions that either detoxify them or activate them to reactive species. The balance between activating and detoxifying enzyme systems governs the rate of delivery of bioactive metabolites to the macromolecular site (Harris, 1991). The resulting macromolecular interaction could be a DNA adduct for carcinogens that are initiating agents or receptor occupancy for chemicals that are tumour promotors. Certain of the DNA adducts produced by such interactions are pro-mutagenic, and replication of the damaged DNA could lead to DNA sequence changes which may result in altered gene expression or mutated gene products. Weisburger & Williams (1981) have suggested that chemical carcinogens be classified as those that interact with DNA (genotoxic) and those that do not (epigenetic or non-genotoxic). The importance of mitotic activity in the latter group has recently been elaborated further (Cohen & Ellwein, 1990).

The implications for invoking the "mechanistic" approach to the selection of appropriate biomarkers are significant. For example, chemicals that stimulate cell proliferation via mitogenesis or cytotoxicity (and subsequent proliferation) might require different biomarkers than chemicals whose major mechanism of action is based on DNA reactivity. In the latter case, measurements of DNA adducts or chromosome alterations may serve as suitable biomarkers, whereas in the former case alternate biomarkers (e.g., cell turnover measurements) may be more appropriate.

## 7.1 Analysis of chemicals and metabolites

Urinary or blood concentrations of several chemicals shown or suspected to be carcinogenic to humans (e.g., arsenic, cadmium, chromium, nickel, benzene, MOCA, polychlorinated biphenyls, styrene, tetrachloroethylene) have for long been used as biomarkers of exposure. Among people exposed to arsenic in a copper smelter, a dose relationship has been observed between the cumulative urinary excretion of arsenic and the risk of lung cancer

(Enterline & Marsh, 1982). For other carcinogenic chemicals, such data are not available, and the measured concentrations may only be interpreted in terms of exposure.

Sensitive techniques based on physicochemical or immunochemical methods for the detection of a variety of carcinogen-modified DNA bases have been developed (Shuker & Farmer, 1992). These include the alkylated purines, aflatoxin-guanine adducts, cis-platinum adducts, thymine glycol, 8-hydroxydeoxyguanosine and PAH-derived adducts.

The aflatoxin marker has been extensively used in both animal and human studies on the relationship between exposure and liver cancer induction. In an ongoing prospective study in Shanghai, China, Ross et al. (1992) reported that subjects with liver cancer were more likely than controls to have detectable concentrations of any of the known aflatoxin metabolites in their urine. Groopman et al. (1991) recently explored the relationship between dietary aflatoxin and excretion in the urine of aflatoxin metabolites and an aflatoxin-DNA adduct. This study was conducted on people living in the Guangxi Autonomous Region, China. These investigators found a positive correlation between aflatoxin $N^7$-guanine and specific metabolites excreted in urine and aflatoxin B1 intake from the previous day.

Exposure to chemical compounds capable of interacting with cellular macromolecules can originate from both exogenous and endogenous sources. Nitrite, nitrate and nitrosating agents can be synthesized endogenously in reactions mediated by bacteria and activated macrophage. In this way endogenous formation of *N*-nitroso compounds can occur at various sites in the body. Endogenously formed *N*-nitroso compounds may be considered as biomarkers of susceptibility; they have been associated in humans with increased risk of cancer of the stomach, oesophagus and urinary bladder, although unequivocal epidemiological data are lacking (Bartsch & Montesano, 1984). The quantitative estimation of endogenous nitrosation in humans can be measured using the *N*-nitroso-proline test. L-proline is utilized as a probe for nitrosatable amines and *N*-nitroso-proline excreted in the urine is determined as a marker. This assay has been applied in some population studies (Bartsch et al. 1991).

## 7.2 Biomarkers for genotoxic carcinogens

### 7.2.1 DNA adducts - general considerations

DNA adducts are being used both as molecular dosimeters (biomarker of exposure) and to assess the genotoxic potential of chemicals (biomarker of effect). The biological significance of such adducts must be assessed on the basis of adduct heterogeneity and of cell and tissue specificity for adduct formation, persistence and repair. Some DNA adducts result in mutation whereas others do not. Mutational specificity in the p53 gene produced by a variety of chemical carcinogens provides evidence that DNA adduct location influences site-specific mutations (Hollstein et al., 1991). Some DNA sequence changes may lead to phenotypic alterations that can be selected, whereas others may not (Compton et al., 1991).

Most tissues are comprised of multiple cell types, and cell types vary considerably in their capacity to convert chemicals to DNA reactive species. For example, lung is composed of multiple cell types in which the relative concentrations of various P-450 isoenzymes and enzymes depends on the cell type. Thus, one compound may produce high concentrations of pro-mutagenic adducts in one cell type, but not in another, whereas the opposite might occur for a compound which is activated by a different P-450 isozyme. DNA adduct concentrations derived from a whole tissue homogenate may grossly overestimate or underestimate adduct concentrations in a given cell type.

Some DNA adducts are repaired quickly, others hardly at all, the adduct loss correlating with cell turnover. Therefore, the concentration and gene location of DNA adducts will change with time after exposure to a genotoxic chemical. Furthermore, the existence of non-random repair in the genome makes it difficult to utilize total DNA repair capacity as an indicator of cell susceptibility to carcinogens (Hanawalt, 1987). It is especially important in human studies to know the duration and timing of exposure for proper evaluation of the biological significance of a given adduct concentration.

Many of the human studies described below have involved measuring metabolism, DNA adduct formation and repair in whole tissues. Techniques need to be refined in order that cell-type-specific variations can be monitored in human tissues, as well as experimental studies in animal models using immunohistochemical techniques for the cell type. Specific localization of DNA adducts has clearly demonstrated that such variations occur. Treatment of rats with the tobacco-specific nitrosamine, 4-($N$-methyl-$N$-nitrosamine)-1-(3-pyridyl)-1-butanone (NNK), results in the induction of tumours in the nasal cavity, lung, liver and pancreas (Hoffman et al., 1984; Rivenson et al., 1988). At low doses of NNK, the prevalence of malignant lung tumours was higher than that observed in other tissues. Cell-type-specific differences have been observed within the lung, the highest concentration of $O^6$-methylguanine having been found in the Clara cells. These cells have the highest levels of the P-450 metabolizing enzymes for NNK and low levels of the $O^6$-methylguanine DNA methyltransferase repair enzymes (Belinsky et al., 1987). Pulmonary tumours are also induced in mice and hamsters following either short- or long-term exposure to this carcinogen (Hecht et al., 1983).

As animals age, DNA adducts are detected in increasing amounts, and, although the relationship of these adducts to tumour development is unclear, they are believed to be derived from dietary constituents or endogenous chemicals such as hormones (Randerath & Randerath, 1991).

### 7.2.2 DNA adducts in human samples

In human studies, it is difficult to obtain non-tumourous target tissue for the quantification of DNA adducts. Lymphocytes are a readily accessible source of human cells that are known to contain DNA adducts. However, there is little information on the reliability of using lymphocyte adduct concentrations for the estimation of target cell or tissue adduct concentrations (Lucier & Thompson, 1987).

Evaluation of dose-response relationships for chemical carcinogens in humans is more complex than in animal models. Radio-labelled carcinogens cannot be administered and the accessibility of tissues and cells is limited. Several approaches to detect DNA adducts in human samples have been evaluated (Wogan & Gorelick, 1985; Santella, 1988). The most frequently used methods are immunoassays and $^{32}$P-postlabelling. Other analytical techniques such as GC-MS and synchronous fluorescence spectroscopy are being used to measure DNA adducts (Weston & Bowman, 1991). In general, immunoassays are both specific and sensitive for alkylated adducts and aflatoxin adducts (Wild & Montesano, 1991; Groopman et al., 1991). However, these methods are not easily applied to quantification of adducts for bulky aromatic hydrocarbons such as benzo[a]pyrene-derived adducts. The main problem is the lack of specificity of the antibodies used in the assay which cross-react with a number of PAH-related adducts (Santella et al., 1985).

The second assay frequently used to quantify DNA adducts in humans is the $^{32}$P postlabelling technique. For a complete description of this assay, see Randerath & Randerath (1991), Beach & Gupta (1992), IARC (1992). The assay is extraordinarily sensitive, being capable of detecting 1 adduct in $10^{10}$ normal nucleotides when appropriate modifications are made to the procedure. The assay is particularly useful for detecting adducts of non-polar polycyclic aromatic hydrocarbons such as 7,8-diol-9,10-oxide-benzo[a]pyrene deoxyguanosine (BPDE). Some DNA modifications such as alkylated DNA adducts, which cannot be easily detected by this assay due to the limitations of the chromatographic systems, can be quantified using a combined $^{32}$P immunochemical precipitation technique (Kang et al., 1993). Studies using the $^{32}$P labelling or immunological methods have been reviewed by Beach & Gupta (1992) and Wild & Montesano (1991). The groups of chemicals studied include alkylating agents, polycyclic aromatic hydrocarbons (PAHs), heterocyclic PAHs, nitro PAHs, cyclopenta-fused PAHs, aromatic amines, alkylbenzenes, quinones, mycotoxins, chemotherapeutic agents, pesticides and aldehydes.

### 7.2.3 Protein adducts

Ehrenberg and his associates pioneered the use of protein adducts as dose monitors for carcinogen exposure in humans, and this work has been reviewed by Hsia (1991). To date the class of proteins that have been most extensively studied are those found in circulating blood, i.e. haemoglobin and albumin. This is mainly because these proteins are relatively abundant and can be easily isolated for analysis.

Haemoglobin has the unique biological property of having a life span equivalent to that of the erythrocyte, which in humans is approximately 120 days, and therefore adduct levels reflect exposures over several months. In contrast, albumin adducts can only be used for assessing recent exposure because of the faster turnover of albumin (half-life of 20-25 days). Protein adducts can be quantified using chemical methods, e.g., aromatic amine release by acid or basic hydrolysis from haemoglobin followed by derivatization and GC-MS analysis (Farmer, 1991), or immunological techniques (e.g., aflatoxin-albumin adducts, Wild et al., 1990).

During the past few years, several human monitoring studies have demonstrated the usefulness of protein adducts as biomarkers of exposure. Examples of chemicals that have been detected as protein adducts in human studies include ethylene and propylene oxide, aniline, cigarette smoke, aromatic amines such as 4-aminobiphenyl, and aflatoxin (Wogan, 1989; Farmer, 1991). Albumin adducts of aflatoxin $B_1$ have also been used in epidemiological studies of their role in the etiology of hepatocellular carcinoma in man. A significant correlation was observed, at the individual level, between dietary intake and the level of albumin-bound aflatoxin in a chronically exposed population in the Gambia (Wild et al., 1992).

### 7.2.4 Cytogenetic methods

Cytogenetic methods are used as biomarkers of exposure to DNA-damaging agents.

Many studies relating to cytogenetic changes in exposed human populations have been reviewed in a special issue of Mutation Research (Anderson, 1988). A second comprehensive review of more recent studies has also been published (Anderson, 1990), and a further review is in press (Anderson, in press).

All human monitoring studies suffer from variability of baseline frequencies (Carrano & Natarajan, 1988) due to the presence of endogenous (gender, age, medical history, etc.) and exogenous factors (life-style, smoking, drinking, eating habits, etc.). Anderson et al. (1991) have investigated the effect of these changes on baseline variability eight times over a two-year period.

In contrast to studies on radiation, for which a marker (dicentric chromosome) has been identified, studies with chemicals have not yet identified a specific marker chromosome. For radiation a dicentric is a quantitative dosimeter. Therefore, after radiation exposure results can be used on an individual basis and a highly exposed individual removed from the radiation source. Results from chemical exposure studies, however, can only be used on a group basis, due to the lack of a specific marker chromosome.

### 7.2.5 Chromosome damage

Both chromosome and chromatid aberrations are induced in individuals exposed to chemical mutagens. The chromosome aberrations are thought to arise from misrepair of lesions in the $G_0$ stage of circulating lymphocytes as well as from precursor cells in bone marrow and thymus (Carrano & Natarajan, 1988). Chromatid aberrations include chromatid breaks, intrachanges and exchanges, while chromosome aberrations include acentric fragments, dicentric chromosomes and ring chromosomes. Balanced translocation and inversions can also arise and are difficult to quantify without banding analysis. Structural aberrations can be classified as unstable and stable depending on their ability to persist in dividing cell populations. Unstable aberrations consist of rings, acentric fragments and other asymmetrical rearrangements, and will lead to the death of the cell. Stable aberrations consist of balanced translocation inversions and other symmetrical rearrangements which can be transmitted to progeny cells at division. Therefore stable aberrations are more biologically significant than unstable ones and could be involved in the cancer process. Many human carcinogens have been shown to produce chromosome damage in populations exposed to them, although no causal relationship has been demonstrated (Sorsa et al., 1992). Proven human carcinogens for which cytogenetic endpoints have been measured in humans and corresponding animal data are available are listed in Table 2.

In a preliminary report of a prospective study among people whose lymphocytes were assayed for chromosome aberrations and SCE, high rates of chromosome aberrations were observed and appeared to be linked to cancer risk, but the finding was of borderline statistical significance (Sorsa et al., 1990).

### 7.2.6 Sister chromatid exchange

Sister chromatid exchange (SCE) is considered to be a more sensitive, rapid and simple cytogenetic end-point than chromosome aberrations for evaluating the genotoxic potential of a variety of mutagenic and carcinogenic agents. It is also used to detect and differentiate many chromosome fragility diseases that predispose to neoplasia. SCE is a DNA-replication-dependent

Table 2. Proven human carcinogens for which cytogenetic end-points have been measured in humans and corresponding data are available for experimental animals[a]

| Agent/exposure | Cytogenetic findings[a] | | | | | |
| | Humans | | | Animals | | |
| | CA | SCE | MN | CA | SCE | MN |
|---|---|---|---|---|---|---|
| **Human carcinogens (Group 1)** | | | | | | |
| Alcoholic beverages | + | + | | − | + | ? |
| Aluminium production | − | − | | | | + |
| Arsenic and arsenic compounds | ? | ? | | + | | − |
| Asbestos | | ? | | − | | + |
| Azathioprine | ? | − | | + | − | + |
| Benzene | + | | | + | + | + |
| Betel quid with tobacco | | + | + | | | |
| Bis(chloromethyl)ether and chloromethyl methyl ether (technical grade) | (+) | | | − | | |
| 1,4-Butanediol dimethanesulfonate (Myleran) | (+) | + | | + | | + |
| Chlorambucil | ? | + | | + | | |
| Cyclosporin | (+) | | | − | | − |
| Coal-tars | + | | | | | |
| Coke production | | + | | | | |
| Combined oral contraceptives | − | − | | | | |
| Cyclophosphamide | + | + | | + | + | + |
| Hexavalent chromium compounds | + | + | | + | + | + |
| Melphalan | + | + | | + | | |

Table 2 (contd).

| Agent/exposure | Cytogenetic findings[a] | | | | | |
| --- | --- | --- | --- | --- | --- | --- |
| | Humans | | | Animals | | |
| | CA | SCE | MN | CA | SCE | MN |
| 8-Methoxypsoralen plus ultraviolet A radiation | – | – | | | + | – |
| Mineral oils, untreated and mildly treated | + | | | | | |
| Nickel compounds | + | – | | ? | | |
| Painter, occupational exposures as | – | | | | | |
| Radon | + | | | – | | |
| Rubber industry | ? | ? | | | | |
| Tobacco products, smokeless | + | | + | | | |
| Tobacco smoke | + | + | + | | + | + |
| Tris(1-aziridinyl)phosphine sulfide (Thiotepa) | (+) | | | + | + | + |
| Vinyl chloride | + | ? | | + | + | + |

[a] From: Sorsa et al. (1992); CA = chromosome aberrations; SCE = sister chromatid exchange; MN = micronuclei
+ = positive result; – = negative result; (+) = equivocal result; ? = doubtful result

phenomenon. Cellular factors such as nucleotide pools, repair and replication enzymes, and biorhythms can play an important role in its formation. A major source of variation can be attributed to the concentration of bromodeoxyuridine relative to the number of lymphocytes in the culture (Das, 1988; Morris, 1991; Morris et al., 1992). In a prospective cancer study (Sorsa et al., 1990), no relationship was observed between the frequency of SCE and the risk of cancer.

### 7.2.7 Micronuclei

Micronuclei are formed by condensation of acentric chromosomal fragments or by whole chromosomes that are left behind during anaphase movements (lagging chromosomes). The presence of micronuclei can therefore be taken as an indication of the previous existence of chromosomal aberrations. To visualize micronuclei, cells have to undergo mitosis. In peripheral lymphocyte cultures it is not easy to distinguish interface nuclei that have undergone a division from those that have not. This makes it difficult to quantify the frequencies of micronuclei for comparative purposes. A method using cytochalasin B can distinguish nuclei that have divided once (French & Morley, 1985).

### 7.2.8 Aneuploidy

Aneuploidy is a condition in which the number of chromosomes in cells of individuals is not an exact multiple of the typical haploid set for the species. Trisomy results when a single extra chromosome is added to a pair of homologous chromosomes. If one chromosome of a pair is missing, the result is monosomy. Absence of the pair is nullisomy. Two or more copies of a homologue result in tetrasomy or polysomy. Cells of individuals with missing or extra chromosomes are hypoploid or hyperploid (UK DH, 1989). The best-known numerical abnormalities result in the syndromes of Down (trisomy of chromosome 21), Klinefelter (sex chromosome genotype is XXY), and Turner (sex chromosome genotype is X0). Aneuploidy is almost always found in human cancers (Dellarco et al., 1985).

### 7.2.9 Mutation

Current somatic gene mutation assays used as biomarkers in human studies select for a change or loss of a normal protein produced by specific genes. Mutations at the X-linked hypoxanthine guanine phosphoribosyl transferase gene in cloned

T-lymphocytes and in the autosomal locus for human leucocyte antigen-A (HLA-A) have provided information on frequency of mutation and molecular spectra of mutants. Detection of haemoglobin variants and loss of the cell surface glycoprotein glycophorin A measured in red blood cells have such limitations that analysis of the DNA mutations induced cannot be made (Compton et al., 1991). The background frequency of each of these specific locus assays varies greatly (Lambert, 1992) and is dependent on numerous confounding factors (e.g., age and smoking). Information on the mutation spectra at a particular locus will be extremely useful in elucidating the mechanisms by which mutations occur in human cells *in vivo*. By comparing spontaneous and chemically induced mutational spectra in different populations, the etiological contributions of both exogenous and endogenous factors to human carcinogenesis could be assessed.

An alternative approach for the measurement of induced base changes which does not require prior selection of a mutant population uses the restriction site mutation technique. This is based on the detection of DNA sequences resistant to the cutting action of specific restriction enzymes, and the amplification of these resistant sequences using the polymerase chain reaction. It may theoretically be applied to the study of DNA base changes in any gene for which the sequence has been determined (Parry et al., 1990).

More relevant biomarkers for chemically induced cancers would, however, preferably measure changes in genes thought to be important for cancer. Mutations that activate proto-oncogenes, which stimulate growth or inactivate suppressor genes to liberate cells from growth constraints, could lead to unregulated proliferation of cancer cells (Weinberg, 1991). For the most part, mutations in oncogenes and tumour suppressor genes have been characterized in tumour tissue. It remains to be determined whether the detection of mutant cells against a background of normal cells can be achieved prior to clinical diagnosis of cancer. Activated oncogenes have already been identified in many human cancers, and considerable progress has been made in elucidating the potential role of chemical carcinogens in the activation of oncogenes and the contribution of the latter to tumorigenesis in animal models (Balmain & Brown, 1988). Brandt-Rauf (1991) presented data from pilot studies that demonstrated the presence of the p21 protein product of the ras oncogene in the serum of 15 out of 18 lung cancer patients who were all current or former smokers. The protein was not found in the serum of any of the 18 healthy non-smoking controls, but was present in 2 out of 8 clinically healthy smoking controls. However, in another study, p21 protein was not detected in 20 smokers in a normal population or in 20 male healthy non-smokers (Brinkworth et al., 1992). The mutational spectrum for the tumour suppressor gene p53 in human tumours has been reviewed by Hollstein et al. (1991). Mutations of the p53 gene are the most common cancer-related genetic changes at the gene level and are widespread over the conserved codons of the p53 gene. Hence, mutational spectra could be compared for tumours at different sites and arising from different etiological backgrounds. The mutational spectrum appears to differ among cancers of the colon, lung, oesophagus, breast, liver, brain, reticuloendothelial tissues and haemopoietic tissues. In two populations where aflatoxin $B_1$ exposure was implicated as one of the etiological factors in hepatocellular carcinomas, the same mutational hotspot (i.e. G-T transversion at codon 249) in the p53 gene has been identified (Hsu et al., 1991).

## 7.3 Biomarkers for non-genotoxic carcinogenesis

Although only a few non-genotoxic human carcinogens have been recognized (e.g., cyclosporin, diethylstilbestrol and estrogenic hormones), many non-genotoxic carcinogens have been identified in rodents. A compilation of NTP rodent data, designed to test the concordance between short-term tests and *in vivo* carcinogenicity assays, showed that more than 30% of rodent carcinogens do not test positively for genotoxicity (Ashby & Tennant, 1991).

The mechanisms of action for non-genotoxic carcinogens need to be considered in predicting human risk from chemical exposures. Although the modes of action of non-genotoxic carcinogens are poorly understood, several have been proposed, including immunosuppression, hormonal effects, promotion, inorganic carcinogenesis, co-carcinogenic effects and solid-state carcinogenesis (Weisburger & Williams, 1981). Recently some of these mechanisms were grouped under the headings of cytotoxicity and mitogenic growth stimulation (Butterworth et al., 1992). Non-genotoxic carcinogens are believed to exert their carcinogenic effects through mechanisms that do not involve direct binding of the chemical or its metabolites to DNA (UK DH, 1989). The key mechanism of non-genotoxic chemicals is to increase cell proliferation, either by mitogenesis of the target cells or by cytotoxicity, which is followed by regenerative cell proliferation (Ramel, 1992). Cohen & Ellwein (1990) suggested that non-genotoxic chemicals can be further categorized as to whether or not their main mechanism of action is mediated via receptor-binding (e.g., dioxin).

Cell replication and proliferation are potential biomarkers of effect. Cell replication is the production of daughter cells by the process of replicative DNA synthesis, while cell proliferation is the enhanced replication of a selected cell population as observed in regenerating tissues. Cells undergoing replicative DNA synthesis (S-phase) are the most commonly used markers. The detection of cell proliferation involves the incorporation of DNA precursors like $^3$H-thymidine or the base analogue 5-bromo-2′-deoxyuridine (BrdU) into cellular DNA during S-phase. This is accomplished by administering these precursors to animals by injection or through implanted osmotic pumps. These S-phase cells can be identified histoautotoradiographically or immuno-histochemically (Goldsworthy et al., 1991). The invasiveness of these techniques currently limits their use to animal studies.

The identification of biomarkers of effect for non-genotoxic carcinogens, whose major mechanism of action is via receptor occupancy, may be difficult. This is primarily because carcinogens of this type activate a variety of genes, some of which may not be involved in the carcinogenic process. However, a good example of a non-genotoxic carcinogen for which there is a good biomarker of effect is 2,3,7,8-tetrachlorinated dibenzo-*p*-dioxin (TCDD). TCDD interacts with a cytosolic receptor that is specific for it and its structural analogues (Poland et al., 1976). In addition to activating a number of growth factor and growth factor receptor genes, TCDD induces a number of enzymes, one of which is cytochrome P-450 1A1. Although the induction of this enzyme is probably not directly related to the biological mechanism leading to cancer from TCDD exposure, it is a sensitive marker for exposure (Tritscher et al., 1992).

# 8. BIOMARKERS OF SUSCEPTIBILITY

This chapter focuses on the genetic predisposition of an individual as it affects susceptibility to chemical materials. There are a number of external factors, such as age, diet and health status, that can also influence the susceptibility of an individual exposed to chemicals. Some discussion will be directed towards the effects of previous exposure on subsequent susceptibility, such as to sensitization and enzyme induction/inhibition by previous exposure. Table 3 lists some genetic and acquired factors affecting susceptibility (Calabrese, 1986).

Although individuals may experience similar environmental exposures, genetic differences in metabolism may produce markedly different doses at the target site and thus a different level of response. Even when target doses are similar, markedly different responses may be noted in individuals due to varying degrees of inherent biological responsiveness. Biomarkers of susceptibility may reflect the acquired or genetic factors that influence the response to exposure. These are pre-existing factors and are independent of the exposure. They are predominantly genetic in origin, although disease, physiological changes, medication and exposure to other environmental agents may also alter individual susceptibility. Biomarkers of susceptibility identify those individuals in a population who have an acquired or genetic difference in susceptibility to the effects of chemical exposure.

Biomarkers of susceptibility indicate which factors may increase or decrease an individual's risk of developing a toxic response following exposure to an environmental agent. Polymorphism is present for some metabolic activation/ deactivation enzymes, including cytochrome P-450 isozymes (Nebert, 1988a, 1988b) and at least one form of glutathione transferase (Seidegard et al., 1990). Differing rates of enzyme activity controlling the activation or detoxification of xenobiotics lead to differences in susceptibility by increasing or decreasing the biologically effective dose of the environmental agent.

The effect may vary between ethnic groups. For instance, there are approximately equal numbers of fast and slow acetylator phenotypes in a Caucasian population, whereas in a Japanese population the distribution is 10% slow acetylators and 90% fast.

Table 3. Some examples of established and suspected biomarkers of susceptibility[a]

| Biomarker of susceptibility | Environmental agent | Disease |
|---|---|---|
| **Genetic** | | |
| Debrisoquine hydroxylation phenotype | cigarette smoke | lung cancer |
| Acetylator phenotype | aflatoxin, | liver cancer, |
| | aromatic amines | bladder cancer |
| Ataxia telangiectasia genotype | bleomycin, epoxides | cancer at various sites |
| Xeroderma pigmentosum genotype | agents that cause oxidative damage to DNA, PAH, aromatic amines and aflatoxin $B_1$ | skin cancer, other cancers |
| Arylhydrocarbon hydroxylase inducibility | polycyclic aromatic hydrocarbons | lung cancer |
| α-1-antitrypsin | cigarette smoke | pulmonary emphysema |
| Franconi's anaemia phenotype | cross-linking agents | acute leukaemia |
| Glucose-6P-dehydrogenese deficiency phenotype | oxidative agents, aromatic amines, nitro-aromatic compounds | poor resistance to oxidative stress, aromatic amines |
| Sickle cell phenotype | aromatic amino and nitro compounds, carbon monoxide, cyanide | anaemia |
| Thalassemia phenotype | lead, benzene | anaemia |
| Erythrocyte porphyria | chloroquine, hexachlorobenzene, lead, various drugs including barbituates, sulfonamides, others | anaemia |

Table 3 (contd).

| Biomarker of susceptibility | Environmental agent | Disease |
|---|---|---|
| **Genetic (contd)** | | |
| Sulfite oxidase deficiency heterozygotes | sulfite, bisulfite, sulfur dioxide | pulmonary disease |
| Alcohol dehydrogenase variant | metabolize alcohols (e.g., ethanol) more quickly than normal | |
| GST$\mu$ phenotype | cigarette smoke | lung cancer |
| Pseudocholinesterase variants | organophosphate and carbamate insecticides, muscle relaxant drugs | neurotoxicity |
| IgA deficiency | respiratory irritants | irritation of respiratory tract |
| phenyl ketones in urine | precursors of phenyl ketones | phenylketonuria |
| **Acquired** | | |
| Deficient diet | chemical | decreased resistance to effects of many chemicals |
| Induced P-450 IIE1 | alcohol consumption | cancer at various sites |
| Antigen-specific antibodies | chemicals, dusts | decreased pulmonary functions, skin rashes |

*a* Modified from Calabrese (1986)

Genetic polymorphisms for drug metabolism have been widely studied using phenotypic assays which involves measuring drug clearance in individuals. Differential rates of metabolism will affect the distribution and persistence of metabolites, which may have implications for the site of toxicity. Epidemiological studies suggest that, with respect to aromatic amines, slow acetylators are more likely to contract bladder cancer but are at *decreased* risk for colo-rectal cancer (Guengerich, 1991; Kadlubar et al., 1992). Polymorphism of N-oxidation has been linked to susceptibility to colonic cancer (Kadlubar et al., 1992) and polymorphism in glutathione *S*-transferase to increased lung cancer, particularly adenocarcinoma (Seidegard et al., 1990).

The methodology for determining the phenotypes of individuals for polymorphisms in metabolizing genes requires the administration of a relevant test drug to the person and the subsequent measurement of its clearance from the body. More recently, techniques based on polymerase chain reactions, using DNA isolated from lymphocytes and other cells, have been developed which allow the detection of genetypes of known polymorphisms involving a variety of xenobiotic-metabolizing enzymes, including GST1 (gluthione-*S*-transferase $\mu$) and NAT2(*N*-acetyl transferase), as well as two cytochrome P-450 isoenzymes: CYP1A1 and CYP2D6 (Bell, 1991; Blum et al., 1991; Wolf et al., 1992; Hirvonen et al., 1992; Hollstein et al., 1992).

Cigarette smoking provides another example that illustrates the effect of genetic polymorphisms on the response to chemicals. Cigarette smoking is associated with the development of lung cancer but not all smokers get lung cancer. This appears to be due in part to genetic variations in arylhydrocarbon hydroxylase activity, which results in a large variability in the binding of benz[*a*]pyrene to DNA in cigarette smokers. A genetically based low level of alpha-1-anti-trypsin activity greatly increases the risk of emphysema from cigarette smoking.

In some situations a genetic trait may make an individual more susceptible to one environmental agent but less so to another. For example, the sickle cell trait predisposes to anaemia and altitude sickness but offers some protection to the individual from infection by the malaria parasite. Inherent differences in susceptibility depend upon variations in the function of genes controlling enzyme activity or the production of other proteins. Although a genotoxic agent may reach the target tissue, the significance of any chromosomal break will depend on the efficiency of the DNA repair mechanisms. In xeroderma pigmentosum the individual is at an increased risk of skin cancer after exposure to UV light because of an inherited defect in DNA repair proteins (Cleaver, 1969). Heterozygotes also have an increased risk of cancer and so the frequency of the gene may affect the incidence of this cancer. Other inherited diseases (e.g., ataxia-telangiectasia) that affect the efficiency of DNA replication or repair may affect susceptibility to carcinogenic agents (Swift et al., 1992). UV-DNA repair capacity has been found to be lower in the lymphocytes isolated from individuals with basal cell and squamous cell carcinoma than in the case of their age-matched controls, and this repair capacity has been found to decrease with increasing age in both groups (Wei et al., 1993).

Another form of susceptibility has an immunological basis. Prior exposure to a chemical may induce an immune response that sensitizes the individual to subsequent exposures. Such responses occur in only a small fraction of the exposed population; an example is the development of pulmonary hypersensitivity to industrial agents such as toluene diisocyanate or cotton dust. The biomarkers of susceptibility are the antigen-specific antibodies developed against the chemical.

# 9. SUMMARY

The Task Group considered biomarkers in three categories, biomarkers of exposure, of effect, and of susceptibility, while recognising that clear distinction of category is often not possible.

The Task Group agreed that the use of biomarkers can improve, and should be used in, the process for the assessment of human health risks caused by exposure to chemicals. Biomarkers may be applied to the estimation of exposure and internal dose in individuals and in groups and may allow identification of those at greater or lesser risk than average.

Biomarkers must be validated before application in the risk assessment process, i.e. the relationship between the biomarker, the exposure, and the health outcome must be established. The selection, validation and application of any biomarker is a complicated process, which will vary for different markers. Examples were selected to illustrate the concepts and principles.

Research and use of biomarkers involves complex ethical, social and legal issues, which may vary in different countries. These issues may impose constraints on research and use of biomarkers in risk assessment and risk management decisions. The ethical, social and legal aspects of biomarkers require careful consideration prior to any application.

# 10. RECOMMENDATIONS

In making the following recommendations, the Task Group recognized the role given IPCS to facilitate and increase coordination of international activities in order to promote the further work needed to define human health effects associated with exposure to chemicals and to provide the basis for priority-setting actions in order to protect public health.

## 10.1 General

- To promote the wider use of validated biomarkers in the risk-assessment process

- To promote interdisciplinary cooperation and communication in order to facilitate application of research findings

- To examine the feasibility of developing a data bank of information on biomarkers applied to the process of risk and their applications

## 10.2 Research

- To develop, refine and validate models to relate biomarkers of exposure and of effect, qualitatively and quantitatively, to exposure and to health outcome, particularly for end-points other than cancer

- To identify and validate biomarkers of susceptibility in relation to the chemical, and to inter-individual variation in response, and investigate genetic polymorphism as a basis for individual hypersensitivity

- To assess the use of information on individual susceptibility in relation to protection of health with due respect to the ethical, social and legal issues

- To develop strategies to link exposure and internal dose with human health outcome by integration of mechanistically validated biomarkers of exposure, effect and susceptibility

## 10.3 Applications

- To develop a practical protocol for use of biomarkers in human studies, taking into account scientific, emotional, ethical, legal and social aspects and including guidelines for risk communication, with emphasis on the right of participants to non-biased, intelligible information

- To encourage the production of certified reference materials for biomarker analyses and support the functioning of international quality assurance programmes

- To include consideration of biomarkers of exposure, effect, and susceptibility in future Environmental Health Criteria monographs

- To consider the need to update this monograph on principles and concepts of biomarkers at an early date

## REFERENCES

ACGIH (1992) 1992-1993 Threshold limit values for chemical substances and physical agents and biological exposure indices. Cincinnati, Ohio, American Conference of Governmental Industrial Hygienists.

Adams RM & Fisher T (1990) Diagnostic patch testing.In: Adams R ed. Occupational skin disease, 2nd ed. Philadelphia, Pennsylvania, W.B. Saunders Company, pp 223-253.

Aitio A (1981) Laboratory quality control. Copenhagen, World Health Organization, Regional Office for Europe, 74 pp (Health Aspects of Chemical Safety - Interim Document 4).

Alessio L, Berlin A, & Roi R (1983) Human biological monitoring of industrial chemicals series. Luxembourg, Office for Official Publications of the Commission of European Communities (Industrial Health and Safety Series).

Alessio L, Berlin A, & Roi R (1984) Biological indicators for the assessment of human exposure to industrial chemicals. Luxembourg, Office for Official Publications of the Commission of the European Communities (Industrial Health and Safety Series - EUR 8903 EN).

Alessio L, Berlin A, & Roi R (1986) Biological indicators for the assessment of human exposure to industrial chemicals. Luxembourg, Office for Official Publications of the Commission of the European Communities (Industrial Health and Safety Series - EUR 10704 EN).

Alessio L, Berlin A, & Roi R (1987) Biological indicators for the assessment of human exposure to industrial chemicals. Luxembourg, Office for Official Publications of the Commission of the European Communities (Industrial Health and Safety Series - EUR 11135 EN).

Alessio L, Berlin A, & Roi R (1988) Biological indicators for the assessment of human exposure to industrial chemicals. Luxembourg, Office for Official Publications of the Commission of the European Communities (Industrial Health and Safety Series - EUR 11478 EN).

Alessio L, Berlin A, & Roi R (1989) Biological indicators for the assessment of human exposure to industrial chemicals. Luxembourg, Office for Official Publications of the Commission of the European Communities (Industrial Health and Safety Series - EUR 12174 EN).

Anderson D ed. (1988) Genetic toxicology testing and biomonitoring of environmental or occupational exposure: human monitoring. Mutat Res, 204(Spec Issue 3): 353-551.

Anderson D ed. (1990) Fundamental and molecular mechanisms of mutagenesis. Mutat Res, 229(Spec Issue 2): 103-247.

Anderson D ed. (in press) Human monitoring II. Mutat Res (Spec Issue).

Anderson D, Francis AJ, Godbert P, Jenkinson PC, & Butterworth KR (1991) Chromosomal aberrations (CA), sister-chromatid exchanges (SCE) and mitogen-induced blastogenesis in cultures of peripheral lymphocytes from 48 control individuals sampled 8 times over 2 years. Mutat Res, 250: 467-476.

Ashby J & Tennant RW (1991) Definitive relationships among chemical structure, carcinogenicity and mutagenicity for 301 chemicals tested by the US National Toxicology Program. Mutat Res, 257: 229-306.

Balmain A & Brown K (1988) Oncogene activation in chemical carcinogenesis. Adv Cancer Res, 51: 147-182.

Barbin A, Laib RJ, & Bartsch H (1985) Lack of miscoding properties of 7-(2-oxoethyl)guanine, the major vinyl chloride-DNA adduct. Cancer Res, 45: 2440-2444.

Bartsch H & Montesano R (1984) Commentary: relevance of nitrosamines to human cancer. Carcinogenesis, 5: 1381-1393.

Bartsch H, Ohshima H, & Shuker DEG (1991) Noninvasive methods for measuring exposure to alkylating agents: recent studies on human subjects. In: Groopman JD & Skipper PL ed. Molecular dosimetry and human cancer - Analytical, epidemiological and social considerations. Boca Raton, Florida, CRC Press, pp 281-301.

Beach AC & Gupta RC (1992) Human biomonitoring and the [32]P-postlabeling assay. Carcinogenesis, 13: 1053-1074.

Bekes JG, Roboz JP, Fischbein A, & Selikoff IJ (1987) Clinical immunology studies in individuals exposed to environmental chemicals. In: Berlin A, Dean J, Draper MH, Smith EMB, & Spreafico F ed. Immunotoxicology. Proceedings of the International Seminar on the Immunological System as a Target for Toxic Damage: Present Status, Open Problems and Future Perspectives. Dordrecht, Boston, Lancaster, Martinus Nijhoff Publishers, pp 347-361 (IPCS Joint Seminar 10; EUR 10041 EN).

Belinsky SA, White CM, Devereux TR, Swenberg JA, & Anderson MW (1987) Cell selective alkylation of DNA in rat lung following low dose exposure to the tobacco specific carcinogen 4-(N-methyl-N-nitrosamino)-1-(3-pyridyl)-1-butanone. Cancer Res, 47: 1143-1148.

Bell DA (1991) Detection of DNA sequence polymorphisms in carcinogen metabolism genes by polymerase chain reaction. Environ Mol Mutagen, 18: 245-248.

Blum M, Demierre A, Grant DM, Helm UA, & Meuer UA (1991) Molecular mechanisms of slow acetylation of drugs and carcinogens in humans. Proc Natl Acad Sci (USA), 88: 5237-5241.

Bond JA, Wallace LA, Osterman-Golkar S, Lucier GW, Buckpitt A, & Henderson RF (1992) Assessment of exposure to pulmonary toxicants: use of biological markers. Fundam Appl Toxicol, 18: 161-174.

Brandt-Rauf PW (1991) Advances in cancer biomarkers as applied to chemical exposures: the ras oncogene and p21 protein and pulmonary carcinogenesis. J Occup Med, 33: 951-955.

Brinkworth MH, Yardley-Jones A, Edwards AJ, Hughes JA, & Anderson D (1992) A comparison of smokers with respect to oncogene products and cytogenetic parameters. J Occup Med, 34: 1181-1189.

Butterworth BE, Popp JA, Conolly RB, & Goldsworthy TL (1992) Chemically induced cell proliferation in carcinogenesis. In: Vainio H, Magee PN, McGregor DB, & McMichael AJ ed. Mechanisms of carcinogenesis in risk identification. Lyon, International Agency for Research on Cancer, pp 279-305 (IARC Scientific Publications No. 116).

Calabrese EJ (1986) Ecogenetics: historical foundation and current status (genetic factors). J Occup Med, **28**: 1096-1102.

Cantin AM, Hubbard RC, & Crystal RG (1989) Glutathione deficiency in the epithelial lining fluid of the lower respiratory tract in idiopathic pulmonary fibrosis. Am Rev Respir Dis, **139**: 370-372.

Cardenas A, Roels H, Bernard AM, Barbon R, Buchet JP, Lauwerys RR, Rosello J, Hotter G, Mutti A, Franchini I, Fels LM, Stolte H, de Broe ME, Nuyts GD, Taylor SA, & Price RG (1993a) Markers of early renal changes induced by industrial pollutants. 1 Application to workers exposed to mercury vapour. Br J Ind Med, **50**: 17-27.

Cardenas A, Roels H, Bernard AM, Barbon R, Buchet JP, Lauwerys RR, Rosello J, Ramis I, Mutti A, Franchini I, Fels LM, Stolte H, de Broe ME, Nuyts GD, Taylor SA, & Price RG (1993b) Markers of early renal changes induced by industrial pollutants. 2 Application to workers exposed to lead. Br J Ind Med, **50**: 28-36.

Carrano AV & Natarajan AT (1988) Considerations for population monitoring using cytogenetic techniques. Mutat Res, **204**: 379-406.

Casanova M, Deyo DF, & Heck HD'A (1989) Covalent binding of inhaled formaldehyde to DNA in the nasal mucosa of Fischer 344 rats: analysis of formaldehyde and DNA by high-performance liquid chromatography and provisional pharmacokinetic interpretation. Fundam Appl Toxicol, **12**: 397-417.

Casanova M, Morgan KT, Steinhagen WH, Everitt JI, Popp JA, & Heck HD'A (1991) Covalent binding of inhaled formaldehyde to DNA in the respiratory tract of rhesus monkeys: pharmacokinetics, rat-to-monkey interspecies scaling, and extrapolation to man. Fundam Appl Toxicol, **17**: 409-428.

Cassito MG, Camerino D, Hanninen H, & Anger KW (1990) International collaboration to evaluate the WHO neurobehavioural core test battery. In: Johnson BL, Anger WK, Durao A, & Xinteras C ed. Advances in neurobehavioural toxicology: applications in environmental and occupational health. Chelsea, Michigan, Lewis Publishers, Inc., pp 203-223.

Costa LG (1992) Effect of neurotoxicants on brain neurochemistry. In: Tilson H & Mitchell C ed. Neurotoxicology. New York, Raven Press Ltd, pp 101-123.

Cavalleri A, Gobba F, Bacchella L, & Ferrari D (1991) Evaluation of serum aminoterminal propeptide of type III procollagen as an early marker of the active fibrotic process in asbestos-exposed workers. Scand J Work Environ Health, **17**: 139-144.

Cleaver JE (1969) Xeroderma pigmentosum: a human disease in which an initial stage of DNA repair is defective. Proc Natl Acad Sci (USA), **63**: 428-435.

Cohen SM & Ellwein LB (1990) Cell proliferation in carcinogenesis. Science, **249**: 1007-1011.

Compton PJE, Hooper K, & Smith MT (1991) Human somatic mutation assays as biomarkers of carcinogenesis. Environ Health Perspect, **94**: 135-141.

Conolly RB & Andersen ME (in press) An approach to mechanism-based cancer risk assessment: formaldehyde. Environ Health Perspect, **101**.

Conolly RB, Monticello TM, Morgan KT, Clewell HJ, & Andersen ME (1992) A biologically-based risk assessment strategy for inhaled formaldehyde. Comments Toxicol, **4**: 269-293.

Das BC (1988) Factors that influence formation of sister chromatid exchanges in human blood lymphocytes. Crit Rev Toxicol, **19**: 43-86.

Dellarco V, Voytek P, & Hollaender A (1985) Aneuploidy: Etiology and mechanisms. New York, London, Plenum Press.

Derosa CT, Choudhury H, & Peirano WB (1991) An integrated exposure/pharmacokinetic based approach to the assessment of complex exposures. Toxicol Ind Health, 7(4): 231-248.

DFG (German Association for the Encouragement of Research) (1992) Commission for the Investigation of Health Hazards of Chemical Compounds in the Work Area: MAK- and BAT-values 1992. Weinheim, VCH.

Droz PO, Wu MM, & Cumberland WG (1989) Variability in biological monitoring of organic solvent exposure. 2 Application of a population physiological model. Br J Ind Med, **46**: 547-558.

Enterline PE & Marsh GM (1982) Cancer among workers exposed to arsenic and other substances in a copper smelter. Am J Epidemiol, **116**: 895-911.

Farmer PB (1991) Analytical approaches for the determination of protein-carcinogen adducts using mass spectrometry. In: Groopman JD & Skipper PL ed. Molecular dosimetry and human cancer - Analytical, epidemiological and social considerations. Boca Raton, Florida, CRC Press, pp 189-210.

FIOH (Työterveyslaitos) (1993) [Biological monitoring. Guide for sample collection 1993.] Helsinki, Finnish Institute of Health, 115 pp (in Finnish).

French M & Morley AA (1985) Measurement of micronuclei in lymphocytes. Mutat Res, **147**: 29-36.

Gibaldi M & Perrier D (1982) Pharmacokinetics. New York, Basel, Marcel Dekker Inc.

Goldsworthy TL, Morgan KT, Popp JA, & Butterworth BE (1991) Guidelines for measuring chemically-induced cell proliferation in specific rat organs. In: Butterworth BE & Sloga TJ ed. Chemically induced cell proliferation: Implications for risk assessment. New York, Wiley-Liss, pp 253-284.

Graham D, Henderson FW, & House D (1988) Neutrophil influx measured in nasal lavages of humans exposed to ozone. Arch Environ Health, **43**: 228-233.

Groopman JD, Sabbioni G, & Wild CP (1991) Molecular dosimetry of human aflatoxin exposures. In: Groopman JD & Skipper PL ed. Molecular dosimetry and human cancer - Analytical, epidemiological and social considerations. Boca Raton, Florida, CRC Press, pp 303-324.

Guengerich FP (1991) Interindividual variation in biotransformation of carcinogens: basis and relevance. In: Groopman JD & Skipper PL ed. Molecular dosimetry and human cancer - Analytical, epidemiological and social considerations. Boca Raton, Florida, CRC Press, pp 27-52.

Hall JA, Saffhill R, Green T, & Hathway DE (1981) The induction of errors during in vitro DNA synthesis following chloroacetaldehyde-treatment of poly(dA-dT) and poly(dC-dG) templates. Carcinogenesis, 2: 141-146.

Hanawalt PC (1987) Preferential DNA repair in expressed genes. Environ Health Perspect, 76: 9-14.

Hancock DG, Moffitt AE, & Hay EB (1984) Hematological findings among workers exposed to benzene at coke oven by-product recovery facility. Arch Environ Health, 39: 414-418.

Harel S, Janoff A, Yu SY, Hurewitz A, & Bergofsky EH (1980) Desmosine radioimmunoassay for measuring elastin degradation *in vivo*. Am Rev Respir Dis, 122: 769-773.

Harris CC (1991) Molecular epidemiology: overview of molecular and biochemical basis. In: Groopman JD & Skipper PL ed. Molecular dosimetry and human cancer - Analytical, epidemiological and social considerations. Boca Raton, Florida, CRC Press, pp 15-26.

Hecht SS, Adams JD, Numoto S, & Hoffmann D (1983) Induction of respiratory tract tumours in Syrian golden hamsters by a single dose of 4-(methylnitrosamino)-1-(3-pyridyl)-1-butanone (NNK) and the effect of smoke inhalation. Carcinogenesis, 4(10): 1287-1290.

Henderson RF (1988) Use of bronchoalveolar lavage to detect lung damage. In: Target organ toxicology: Lung. New York, Raven Press, pp 239-268.

Henderson RF, Bechtold WE, Bond JA, & Sun JD (1989) The use of biological markers in toxicology. Crit Rev Toxicol, 20: 65-82.

Hirvonen A, Husgafvel-Pursiainen K, Karjalainen A, Anttila S, & Vainio H (1992) Point-mutational MspI and Ile-Val polymorphisms closely linked in the CYP1A1 Gene: Lack of association with susceptibility to lung cancer in a Finnish study population. Cancer Epidemiol Biomarkers Prev, 1: 485-489.

Hoffman D, Rivenson A, Amin S, & Hecht SS (1984) Dose-response study of the carcinogenicity of tobacco-specific N-nitrosamines in F344 Rats. J Cancer Res Clin Oncol, 108: 81-86.

Hollstein M, Sidransky D, Vogelstein B, & Harris CC (1991) p53 Mutations in human cancers. Science, 253: 49-53.

Hollstein MC, Wild CP, Bleicher F, Chutimataewin S, Harris CC, Srivatanakul P, & Montesano R (1992) p53 Mutations and aflatoxin $B_1$h exposure in hepatocellular carcinoma patients from Thailand. Int J Cancer, 53: 1-5.

Holsapple MP, Snyder NK, Wood SC, & Morris DL (1991) A review of 2,3,7,8-tetrachlorodibenzo-p-dioxin-induced changes in immunocompetence: 1991 update. Toxicology, 69: 219-255.

Horak F (1985) Manifestation of allergic rhinitis in latent-sensitized patients. A prospective study. Arch Otorhinolaryngol, 242: 224-239.

Hsia MTS (1991) Carcinogen-macromolecular adducts as biomarkers in human cancer risk assessment. Biomed Environ Sci, **4**: 104-112.

Hsu IC, Metcalf RA, Sun T, Welsh J, Wang NJ, & Harris CC (1991) Mutational hotspot in the p53 gene in human hepatocellular carcinomas. Nature (Lond), **350**: 427-428.

IARC (1992) In: Vainio H, Magee PN, McGregor D, & McMichael AJ ed. Mechanisms of carcinogenesis in risk identification. Lyon, International Agency for Research on Cancer (IARC Scientific Publications No. 116).

Jennings AM, Wild G, Ward JD, & Milford Ward A (1988) Immunologic abnormalities 17 years after accidental exposure to 2,3,7,8-tetrachlorodibenzo-p-dioxin. Br J Ind Med, **45**: 701-704.

Kadlubar FF, Butler MA, Kaderlik KR, Chou H-S, & Lang NP (1992) Polymorphisms for aromatic amine metabolism in humans: relevance for human carcinogenesis. Environ Health Perspect, **98**: 69-74.

Kang H-I, Konishi C, Kuroko T, & Huh N-K (1993) A highly sensitive and specific method for quantitation of O-alkylated DNA adducts and its application to the analysis of human tissue. Environ Health Perspect, **99**: 269-271.

Lambert B (1992) Biological markers in exposed humans: gene mutation. In: Vainio H, Magee PN, McGregor DB, & McMichael AJ ed. Mechanisms of carcinogenesis in risk identification. Lyon, International Agency for Research on Cancer, pp 535-542 (IARC Scientific Publications No. 116).

Lamm SH, Walters AS, Wilson R, Byrd DM, & Grunwalt H (1989) Consistencies and inconsistencies underlying the quantitative assessment of leukemia risk from benzene exposure. Environ Health Perspect, **82**: 289-297.

Lassalle P, Cosset P, & Aerts C (1990) Abnormal secretion of interleukin-1 and tumor necrosis factor alpha by alveolar macrophages in coal worker's pneumoconiosis: comparison between simple pneumoconiosis and progressive massive fibrosis. Exp Lung Res, **16**: 73-80.

Lucier G & Thompson CL (1987) Issues in biochemical applications to risk assessment: when can lymphocytes be used as surrogate markers? Environ Health Perspect, **76**: 187-191.

Mattison DR (1991) An overview on biological markers in reproductive and development toxicology: concepts and definitions, and use in risk assessment. Biomed Environ Sci, **4**: 8-34.

Morris SM (1991) The genetic toxicology of 5-bromodeoxyuridine in mammalian cells. Mutat Res, **258**: 161-188.

Morris SM, Domon OE, McGarrity LJ, Kodell RL, & Casciano DA (1992) Effect of bromodeoxyuridine on the proliferation and growth of ethyl methanesulfonate-exposed P3 cells: relationship to the induction of sister chromatid exchanges. Cell Biol Toxicol, **8**: 75-87.

Nebert DW (1988a) The 1986 Bernard B. Brodie award lecture. The genetic regulation of drug-metabolizing enzymes. Drug Metab Dispos, **16**: 1-8.

Nebert DW (1988b) Genes encoding drug metabolizing enzymes: possible role in human disease. Basic Life Sci, **43**: 45-64.

Nielsen J, Welinder H, Horstmann V, & Skerfving S (1992) Allergy to methyltetrahydrophthalic anhydride in epoxy resin workers. Br J Ind Med, **49**: 769-775.

O'Callaghan JP (1991) Quantification of glial fibrillary acidic protein: Comparison of slot-immunobinding assays with a novel sandwich ELISA. Neurotoxicol Teratol, **13**: 275-281.

Parry JM, Shamsher M, & Skibinski DOF (1990) Restriction site mutation analysis, a proposed methodology for the detection and study of DNA base changes following mutation exposure. Mutagenesis, **5**: 209-212.

Passingham BJ, Farmer FB, Bailey E, Brookes AGF, & Yates DW (1988) 2-Hydroxyethylation of haemoglobin in man. In: Methods for detecting DNA and damaging agents in humans: applications in cancer epidemiology and prevention. Lyon, International Agency for Research on Cancer, pp 279-285 (IARC Scientific Publications No. 89).

Piguet PF, Collart MA, Grau GE, Sappino AP, & Vassalli P (1990) Requirement of tumor necrosis factor for development of silica-induced pulmonary fibrosis. Nature (Lond), **344**: 245-247.

Poland A, Glover E, & Kende AS (1976) Stereospecific, high affinity binding of 2,3,7,8-tetrachlorodibenzo-p-dioxin by hepatic cytosol. Evidence that the binding species is receptor for induction of aryl hydrocarbon hydroxylase. J Biol Chem, **251**: 4936-4946.

Rajamaki A (1984) Assessment of early myelotoxicity. In: Aitio A, Riihimaki V, & Vainio H ed. Biological monitoring and surveillance of workers exposed to chemicals. Washington, New York, London, Hemisphere Publishing Corporation, pp 303-308.

Ramel C (1992) Genotoxic and nongenotoxic carcinogens: mechanisms of action and testing strategies. In: Vainio H, Magee PN, McGregor DB, & McMichael AJ ed. Mechanisms of carcinogenesis in risk identification. Lyon, International Agency for Research on Cancer, pp 195-209 (IARC Scientific Publications No. 116).

Ramsay JC & Andersen ME (1984) A physiologically based description of the inhalation pharmacokinetics of styrene monomer in rats and humans. Toxicol Appl Pharmacol, **73**: 159-175.

Randerath K & Randerath E (1991) [32]P-postlabeling analysis of mutagen- and carcinogen-adducts and age-related DNA modifications (I-compounds). In: Groopman JD & Skipper PL ed. Molecular dosimetry and human cancer - Analytical, epidemiological and social considerations. Boca Raton, Florida, CRC Press, pp 131-150.

Reynolds HY (1987) Bronchoalveolar lavage. Am Rev Respir Dis, **135**: 250-263.

Rivenson A, Hoffman D, Prokopszyk B, Amin S, & Hecht SS (1988) Induction of lung and pancreas tumours in F344 rats by tobacco-specific and areca derived N-nitrosamines. Cancer Res, **48**: 6912-6917.

Roels H, Bernard AM, Cardenas AR, Buchet JP, Lauwerys RR, Hotter G, Ramis I, Mutti A, Frandhini I, Bundschuh I, Stolte H, de Broe ME, Nuyts GD, Taylor SA, & Price RG (1993) Markers of early renal changes induced by industrial pollutants. 3 Application to workers exposed to cadmium. Br J Ind Med, **50**: 37-48.

Ross RK, Yuan J-M, Wogan GN, Qian G-S, Tu J-T, Groopman JD, Gao Y-T, & Henderson BE (1992) Urinary aflatoxin biomarkers and risk of hepatocellular carcinoma. Lancet, **339**: 943-946.

Santella RM (1988) Application of new techniques for the detection of carcinogen adducts to human population monitoring. Mutat Res, **205**: 271-282.

Santella RM, Gaspara F, & Hsieh L (1985) Quantitation of carcinogen-DNA adducts with monoclonal antibodies. Prog Exp Tumour Res, **31**: 63-75.

Seidegard J, Pero RW, Markowitz MM, Rousch G, Miller DG, & Beattle EJ (1990) Isoenzyme(s) of glutathione transferase (class mu) as a marker for the susceptibility to lung cancer: A follow-up study. Carcinogenesis, **11**: 33-36.

Seppalainen A-M, Hernberg S, & Kock B (1979) Relationship between blood lead levels and nerve conduction velocities. Neurotoxicology, **1**: 313-332.

Shuker DEG & Farmer PB (1992) Relevance of urinary DNA adducts as markers of carcinogen exposure. Chem Res Toxicol, **5**: 450-460.

Singer B, Chavez SJ, & Kusmierek JT (1987) The vinyl chloride-derived nucleoside $N^2,3$-ethenoguanosine, is a highly efficient mutagen in transcription. Carcinogenesis, **8**: 745-747.

Sorsa M, Ojajarvi A, & Salomaa S (1990) Cytogenetic surveillance of workers exposed to genotoxic chemicals: preliminary experiences from a prospective cancer study in a cytogenetic cohort. Teratog Carcinog Mutagen, **10**: 215-221.

Sorsa M, Wilbourn J, & Vanio H (1992) Human cytogenetic damage as a predictor of cancer risk. In: Vanio H, Magee PN, McGregor D, & McMichael AJ ed. Mechanisms of carcinogenesis in risk identification. Lyon, International Agency for Research on Cancer, pp 543-554 (IARC Scientific Publications No. 116).

Stone PJ, Bryan-Rhadfi J, & Lucey EC (1991) Measurement of urinary desmosine by isotope dilution and high performance liquid chromatography. Am Rev Respir Dis, **144**: 284-290.

Sub-Committee on Skin Tests of the European Academy of Allergology and Clinical Immunology (1989) In: Dreborg S ed. Skin tests for diagnosis of IgE-mediated allergy. Position paper. Allergy, **44**(Suppl 10): 31-37.

Sullivan JB (1989) Immunological alterations and chemical exposure. Clin Toxicol, **27**: 311-343.

Swenberg JA, Fedtke N, Fennell TR, & Walker VE (1990) Relationships between carcinogen exposure, DNA adducts and carcinogenesis. In: Clayson DB, Munro IC, Shubik P, & Swenberg JA ed. Progress in predictive toxicology. Amsterdam, Oxford, New York, Elsevier Science Publishers, pp 161-184.

Swift M, Morrel D, Massey RB, & Chase CL (1992) Incidence of cancer in 161 families affected by ataxia-telangiectasia. New Engl J Med, **325**: 1831-1836.

Townsend JC, Ott MG, & Fishbeck WA (1978) Health exam findings among individuals occupationally exposed to benzene. J Occup Med, **20**: 543-548.

Tritscher AM, Goldstein JA, Portier CJ, McCoy ZM, Clark GC, & Lucier GW (1992) Dose-response relationships for chronic exposure to 2,3,7,8-tetrachlorodibenzo-p-dioxin in a rat tumour promotion model: quantification and immunolocalization of CYP1A1 and CYP1A2 in the liver. Cancer Res, 52: 3436-3442.

UK DH (1989) Committee on Mutagenicity of Chemicals in Food. Guidelines for the testing of chemicals for mutagenicity - Consumer Products and the Environment. London, Department of Health, Her Majesty's Stationery Office (HMSO), 99 pp (Report No. 35).

UK HSE (1991) Guidance on laboratory techniques in occupational medicine, 5th ed. London, Health and Safety Executive, Library and Information Services, 117 pp.

US EPA (1991) Formaldehyde risk assessment update. Washington, DC, US Environmental Protection Office, Office of Toxic Substances.

US EPA (1987) Assessment of health risks to garment workers and certain home residents from exposure to formaldehyde. Washington, DC, US Environmental Protection Office, Office of Toxic Substances.

US NRC (US National Research Council) (1989a) Biologic markers in pulmonary toxicology. Washington, DC, National Academy Press, 179 pp.

US NRC (US National Research Council) (1989b) Biologic markers in reproductive toxicology. Washington, DC, National Academy Press, 395 pp.

US NRC (US National Research Council) (1992a) Biologic markers in immunotoxicology. Washington, DC, National Academy Press, 206 pp.

US NRC (US National Research Council) (1992) Environmental neurotoxicology. Report of Committee on Neurotoxicology and Models for Assessing Risk. Washington, DC, National Academy Press, 154 pp.

Wei Q, Matanoski GM, Farmer ER, Hedayati MA, & Grossman L (1993) DNA repair and aging in basal cell carcinoma: a molecular epidemiology study. Proc Natl Acad Sci (USA), 90: 1614-1618.

Weinberg RA (1991) Tumour suppressor genes. Science, 254: 1138-1146.

Weisburger JH & Williams GM (1981) Carcinogen testing: current problems and new approaches. Science, 214: 401-407.

Weston A & Bowman ED (1991) Fluorescence detection of benzo(a)pyrene-DNA adducts in human lung. Carcinogenesis, 12: 1445-1449.

WHO (1991) IPCS Environmental Health Criteria 119: Principles and methods for the assessment of nephrotoxicity associated with exposure to chemicals. Geneva, World Health Organization, 266 pp.

WHO (1992a) IPCS Environmental Health Criteria 134: Cadmium. Geneva, World Health Organization, 280 pp.

WHO (1992b) IPCS Environmental Health Criteria 141: Quality management for chemical safety testing. Geneva, World Health Organization, 112 pp.

WHO (in press) Principles for assessment of risks from exposure to chemicals (Part A).

Wild CP & Montesano R (1991) Immunological quantitation of human exposure to aflatoxins and N-nitrosamines. In: Vanderlaan M, Stanker LH, Watkins BE, & Roberts DW ed. Immunoassays for trace chemical analysis. Washington, DC, American Chemical Society, pp 215-228 (ACS Symposium Series No. 451).

Wild CP, Jiang YZ, Sabbioni G, Chapot B, & Montesano R (1990) Evaluation of methods for quantitation of aflatoxin-albumin adducts and their application to human exposure assessment. Cancer Res, 50: 245-251.

Wild CP, Hudson GJ, Sabbioni G, Chapot B, Hall AJ, Wogan GN, Whittle H, Montesano R, & Groopman JD (1992) Dietary intake of aflatoxins and the level of albumin-bound aflatoxin in peripheral blood in The Gambia, West Africa. Cancer Epidemiol Biomarkers Prev, 1: 229-234.

Wogan GN (1989) Markers of exposure to carcinogens: methods for human monitoring. J Am Coll Toxicol, 8: 871-881.

Wogan GN & Gorelick NJ (1985) Chemical and biochemical dosimetry of exposure to genotoxic chemicals. Environ Health Perspect, 62: 5-18.

Wolf CR, Dale Smith CA, Gough AC, Moss JE, Vallis KA, Howard G, Carey FJ, Mills K, McNee W, Carmichael J, & Spurr NK (1992) Relationship between the debrisoquine hydroxylase polymorphism and cancer susceptibility. Carcinogenesis, 13: 1035-1038.

Yanagisawa Y, Nishimura H, Matsuki H, Osaka F, & Kasuga H (1986) Personal exposure and health effect relationship for $NO_2$ with urinary hydroxyproline to creatinine ratio as indicator. Arch Environ Health, 41: 41-48.

# RESUME

Le groupe spécial a divisé les marqueurs biologiques en trois catégories, les marqueurs d'exposition, les marqueurs d'effet et les marqueurs de réceptivité, tout en admettant que bien souvent il n'était pas possible d'établir une distinction nette entre ces diverses catégories.

Le groupe spécial a admis que le recours aux marqueurs biologiques pouvait améliorer l'évaluation des risques pour la santé humaine découlant d'une exposition à des produits chimiques et qu'il fallait donc en faire usage. Les marqueurs biologiques peuvent être utilisés pour évaluer l'exposition et la dose interne chez des individus et des groupes et peuvent faciliter l'identification de ceux de ces individus ou de ces groupes qui sont plus ou moins exposés aux risques que la moyenne.

Avant d'utiliser les marqueurs biologiques pour l'évaluation du risque il faut les valider, c'est-à-dire établir la relation qui existe entre le marqueur, l'exposition et ses conséquences pour la santé. Le choix, la validation et l'utilisation de tout marqueur biologique constituent des processus complexes qui varient d'un marqueur à l'autre. Pour mettre en lumière ces notions et ces principes on a choisi un certain nombre d'exemples.

La recherche et l'utilisation des marqueurs biologiques soulèvent des questions complexes sur le plan éthique, social et juridique, qui peuvent d'ailleurs varier d'un pays à l'autre. Il peut s'en suivre un certain nombre de contraintes imposées à la recherche et à l'utilisation des marqueurs biologiques dans l'évaluation des risques et dans les décisions relatives à la gestion de ces risques. Avant toute application il importe d'étudier avec soin les aspects éthiques, sociaux et juridiques des marqueurs biologiques.

# RECOMMANDATIONS

En formulant les recommandations ci-après, le groupe de travail a pris en considération le rôle dévolu au PISC, à savoir de faciliter et de développer la coordination des activités internationales afin d'encourager les travaux à poursuivre pour définir les effets sur la santé humaine qu'entraîne l'exposition aux substances chimiques et de jeter les bases des actions prioritaires à entreprendre pour protéger la santé publique.

## 1. Généralités

● Encourager un plus large recours aux marqueurs biologiques dans l'évaluation des risques

● Favoriser la coopération et la communication interdisciplinaires afin de faciliter l'application des résultats de la recherche

● Etudier la faisabilité d'une banque de données sur les marqueurs biologiques dans l'évaluation des risques et ses applications

## 2. Recherche

● Mettre au point, affiner et valider des modèles qui permettent de corréler les marqueurs biologiques de l'exposition et des effets tant qualitativement que quantitativement, à l'exposition et aux conséquences biologiques, en particulier aux conséquences autres que le cancer.

● Recenser et valider des marqueurs biologiques de réceptivité, par rapport à la réaction aux produits chimiques et aux variations interindividuelles à cette réaction et étudier le polymorphisme génétique en tant que base de l'hypersensibilité individuelle.

● Voir dans quelle mesure il est possible d'utiliser certains renseignements sur la réceptivité individuelle pour la protection de la santé, dans le respect des impératifs éthiques, sociaux et juridiques.

● Mettre au point des stratégies afin de relier l'exposition et la dose interne aux conséquences biologiques pour l'homme grâce à une synthèse de biomarqueurs de l'exposition, de l'effet et de la réceptivité, mécanistiquement validés.

## 3. Applications

● Mettre au point un protocole pratique pour l'utilisation des marqueurs biologiques dans les études sur l'homme, qui prenne en considération les aspects scientifiques, émotionnels, éthiques, juridiques et sociaux et qui comporte des directives en matière de communication, en insistant sur le droit des participants à disposer d'informations intelligibles et non biaisées.

● Encourager la production de substances de référence homologuées pour l'analyse des marqueurs biologiques et aider les programmes internationaux d'assurance de la qualité à fonctionner.

● Faire figurer des considérations sur les biomarqueurs d'exposition, d'effet et de réceptivité dans les futures monographies de la série Critères d'hygiène de l'environnement.

● Examiner s'il est nécessaire de mettre à jour à bref délai la présente monographie consacrée aux principes et conceptions en matière de marqueurs biologiques.

# RESUMEN

El Grupo Especial distinguió tres clases de biomarcadores: de exposición, de efecto y de susceptibilidad, reconociendo sin embargo que a menudo resulta imposible establecer claramente la pertenencia a una de esas clases.

El Grupo Especial coincidió en que los biomarcadores permiten mejorar, y deben emplearse a ese efecto, el proceso de evaluación de los riesgos que para la salud humana conlleva la exposición a productos químicos. Los biomarcadores se pueden emplear para calcular la exposición y la dosis interna recibida por individuos y grupos, con la consiguiente identificación de quienes sufren un mayor o menor riesgo que la media.

Los biomarcadores deben ser validados antes de aplicarlos a la evaluación del riesgo, lo que significa que debe determinarse la relación entre el biomarcador, la exposición y el estado de salud. La selección, validación y empleo de cualquier biomarcador es un proceso complicado, distinto para cada marcador. Se eligieron algunos ejemplos para ilustrar los conceptos y principios relacionados.

La investigación y el empleo de los biomarcadores plantea complejas cuestiones éticas, sociales y jurídicas, que pueden diferir de un país a otro. Algunos de esos problemas limitan el alcance de las investigaciones sobre los biomarcadores y de su aplicación a la evaluación de riesgos y la adopción de decisiones relacionadas con la gestión de riesgos. Los problemas éticos, sociales y jurídicos que plantean los biomarcadores deben ser objeto de un detenido análisis antes de su eventual uso.

# RECOMENDACIONES

Al formular las siguientes recomendaciones, el Grupo Especial reconoció la función asignada al IPCS de facilitar e intensificar la coordinación de las actividades internacionales con miras a fomentar los trabajos que aún será necesario realizar para determinar los efectos sobre la salud relacionados con la exposición a productos químicos, así como para establecer las bases que permitan señalar las prioridades a que haya que atenerse para proteger la salud pública.

## 1. Recomendaciones generales

- Promover un mayor uso de los biomarcadores validados en el proceso de evaluación de riesgos.

- Fomentar la cooperación y la comunicación interdisciplinarias para facilitar la aplicación de los resultados de las investigaciones.

- Estudiar la posibilidad de crear un banco de datos sobre biomarcadores aplicados a la evaluación de riesgos y sus posibles usos.

## 2. Investigaciones

- Desarrollar, perfeccionar y validar modelos aptos para relacionar los biomarcadores de exposición y de efecto, cualitativa y cuantitativamente, con la exposición y con el estado de salud, sobre todo para puntos finales distintos del cáncer.

- Identificar y validar biomarcadores de susceptibilidad en relación con el producto químico y con la variación interindividual de la respuesta, e investigar la influencia del polimorfismo genético en la hipersensibilidad individual.

- Evaluar el uso de la información referente a la susceptibilidad individual desde la perspectiva de una protección de la salud atenta a los aspectos éticos, sociales y jurídicos.

- Formular estrategias para relacionar la exposición y la dosis interna con el estado de salud mediante la integración de biomarcadores de exposición, efecto y susceptibilidad validados sistemáticamente.

## 3. Aplicaciones

- Elaborar un protocolo práctico para el uso de biomarcadores en los estudios realizados en el hombre, teniendo en cuenta los aspectos científicos, psicológicos, éticos, jurídicos y sociales, con inclusión de directrices para la comunicación del riesgo y haciendo hincapié en el derecho de los participantes a una información inteligible e imparcial.

- Fomentar la producción de material de referencia certificado para el análisis de biomarcadores y respaldar la aplicación de programas internacionales de garantía de la calidad.

- Incluir la consideración de los biomarcadores de exposición, efecto y susceptibilidad en las futuras monografías de la serie Criterios de Salud Ambiental.

- Tener presente la necesidad de actualizar con prontitud la presente monografía sobre los principios y nociones relativos a los biomarcadores.

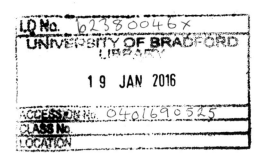